Historic Bridges

Evaluation, Preservation, and Management

Historic Bridges

Bridges

Evaluation, Preservation, and Management

Edited by Hojjat Adeli

CRC Press
Taylor & Francis Group
Boca Raton London New York

CRC Press is an imprint of the
Taylor & Francis Group, an **informa** business

CRC Press
Taylor & Francis Group
6000 Broken Sound Parkway NW, Suite 300
Boca Raton, FL 33487-2742

First issued in paperback 2019

ISBN-13: 978-1-4200-7995-1 (hbk)
ISBN-13: 978-0-367-38747-1 (pbk)

Library of Congress Cataloging-in-Publication Data

Historic Bridges Conference (8th : 2008 : Columbus, Ohio)
  Historic bridges : preservation, rehabilitation, and maintenance / editor, Hojjat Adeli.
      p. cm.
  Includes bibliographical references and index.
  ISBN 978-1-4200-7995-1 (alk. paper)
  1. Historic bridges--Conservation and restoration--United States--Congresses.
I. Adeli, Hojjat, 1950- II. Title.

TG23.H57 2008
624.2028'8--dc22   2008008949

Visit the Taylor & Francis Web site at
http://www.taylorandfrancis.com

and the CRC Press Web site at
http://www.crcpress.com

Contents

PART 1   History

PART 2   Management

PART 3   Evaluation

PART 4   Preservation, Rehabilitation and Restoration

Preface

A good number of engineers, historians, academics, and laypeople are dedicated to the preservation of historic bridges. The Ohio State University started the Historic Bridges Conferences (HBCs) on November 1, 1985 with the encouragement of one of its graduates, Abba Lichtenstein. Over the years additional successful HBCs followed in Columbus, Ohio, March 11, 1988, on October 5, 1990, August 27–29, 1992; in Cincinnati, Oct. 23–24 1997; Wheeling, West Virginia, October 21–23, 1999; and Cleveland, Ohio, September 19–22, 2001.

The 8th Historic Bridges Conference (HBC) was held in Columbus, Ohio, April 28–29, 2008, under the primary and generous sponsorship of Abba Lichtenstein and the Allan King Sloan Foundation. Further, the conference was co-sponsored by The Ohio State University College of Engineering and Department of Civil and Environmental Engineering and Geodetic Science, the Historic Bridge Foundation, Jones-Stuckey Ltd., Mead & Hunt, Inc., and VJM Metal Craftsman, LLC. The contributions of sponsors and co-sponsors, and the support of Dean Bud Baeslack of The Ohio State University are gratefully acknowledged.

The conference consisted of twelve oral presentations and five poster presentations. These papers were selected based on reviews by the following members of the Program Committee: Steven Fenves, Dario Gasparini, Abba Lichtenstein, David A. Simmons, William Vermes, and myself. Further, Abba G. Lichtenstein was the Luncheon Speaker. The title of his lecture was *Saving Historic Bridges: How?* Also, at the icebreaker on April 28, Dario Gasparini gave a short presentation on *Why are 19th Century Stone Arches Relevant to 21st Century Structural Engineers?*

Authors of fourteen papers were asked to revise their papers based on the comments of the reviewers. The revised papers constitute Chapters Two to Fifteen of this book. Professor Steven Fenves was invited to write an introductory chapter, Chapter One in the book.

To bring the book out by the conference date the publishing team at Taylor & Francis, Mr. Joe Clements, Ms. Jill Jurgensen, and Ms. Suzanne Lassandro had to place the book on a fast track for which I am grateful. Finally, I would like to thank Ms. Cindy Sopher who prepared numerous versions of the call-for-papers and conference preliminary and final programs and worked on all the conference logistics.

Hojjat Adeli
The Ohio State University

The Editor

Hojjat Adeli received his Ph.D. in Civil (Structural) Engineering from Stanford University in 1976 after graduating from the University of Tehran, Iran, in 1973. He is Professor of Civil and Environmental Engineering and Geodetic Science and the holder of the Abba G. Lichtenstein Professorship at The Ohio State University. He has authored over 420 research and scientific publications in various fields of computer science, engineering, applied mathematics, and medicine. He has authored 11 books including *Control, Optimization, and Smart Structures — High-Performance Bridges and Buildings of the Future* (Wiley, 1999). His forthcoming book is entitled *Intelligent Infrastructure — Neural Networks, Wavelets, and Chaos Theory for Intelligent Transportation Systems and Smart Structures* (CRC Press/ Taylor & Francis, 2008). He is the Founder and Editor-in-Chief of the international research journals *Computer-Aided Civil and Infrastructure Engineering*, in publication since 1986 and *Integrated Computer-Aided Engineering*, in publication since 1993. He is the quadruple winner of The Ohio State University College of Engineering Lumley Outstanding Research Award. In 1998 he received the *Distinguished Scholar Award*, The Ohio State University's highest research award, *"in recognition of extraordinary accomplishment in research and scholarship"*. In 2005, he was elected Honorary (Distinguished) Member, American Society of Civil Engineers: *"for wide-ranging, exceptional, and pioneering contributions to computing in civil engineering and extraordinary leadership in advancing the use of computing and information technologies in many engineering disciplines throughout the world."* In 2006, he received the ASCE Construction Management Award for *"For development of ingenious computational and mathematical models in the areas of construction scheduling, resource scheduling, and cost estimation,"* In 2007, he received The Ohio State University College of Engineering Peter L. and Clara M. Scott Award for Excellence in Engineering Education as well as the Charles E. MacQuigg Outstanding Teaching Award. He has presented Keynote Lectures at 64 research conferences held in 38 different countries.

Contributors

Charles Birnstiel
Hardesty & Hanover, LLP

Douglas E. Bond
McMullan and Associates

Brian Brenner
Fay Spofford & Thorndike
Tufts University

Stephen Buonopane
Bucknell University

Martin P. Burke, Jr.
M. P. Burke Bridges

Ching Chiaw Choo
California State University, Fresno

Ian Engstrom
Parsons Transportation Group

Steven Fenves
Carnegie Mellon University

Joseph J. Fonzi
Parsons Brinckerhoff, Inc.

Robert M. Frame III
Mead & Hunt Inc.

Issam E. Harik
University of Kentucky

Sean Kelton
Bucknell University

S. D. Daniel Lee
Fay Spofford & Thorndike

Alan J. Lutenegger
University of Massachusetts, Amherst

Denis J. McMullan
McMullan and Associates

Bob Newbery
Wisconsin Department of
Transportation

Steven A. Olson
HNTB Corporation

Kevin L. Rens
University of Colorado, Denver

Frederick R. Rutz
J.R. Harris & Company

Allan King Sloan
Arthur D. Little, International

Amy Squitieri
Mead & Hunt, Inc.

Huan Cheng Tang
Zongtie Major Bridge
Engineering Group

Preston Vineyard
Parsons Brinckerhoff

1 Introduction

Steven J. Fenves

The chapters of this book devoted to historical bridges deal with their history, management, evaluation, preservation, rehabilitation and restoration. This introductory chapter presents the subject of historic bridges as viewed by an engaged observer, rather than an active practitioner or researcher in the field.

I first became exposed to historic bridges as an undergraduate in civil engineering at the University of Illinois. In Civil Engineering Hall every available wall surface in the stairwells, hallways and drafting rooms was covered with large black-and-white photographs of civil engineering works from all over the United States, a good number of them historic bridges. Going to class, on breaks between classes and even during some dull drafting sessions, we students scanned the pictures, admired the works they represented, and tried to figure out how they worked, what made them stand up. My classmates and I felt that these photographs contributed significantly to our evolution into civil engineers.

By the time I got to Carnegie Mellon University in Pittsburgh, I knew many of Pittsburgh's historic bridges from photographs and previous visits. After a boat tour with my family up and down Pittsburgh's fabled Three Rivers, looking up at dozens and dozens of bridges, I decided to replicate the Illinois photograph gallery at Carnegie Mellon University with pictures of Pittsburgh's bridges, guided by the book by White and van Bernewitz.[1] This turned out to be a 28-year journey, through which I became better known in Pittsburgh by my avocation than by my vocation. I met many of the significant bridge people in Pittsburgh and had a range of great experiences: conducting the annual bridge walk for the junior structures class, giving slide presentations at many venues, curating exhibits, serving as assistant to a demolition contractor bringing down a 1884 bridge, narrating boat tours and portions of a television documentary,[2] documenting the Roeblings' daring designs for tripartite bridges at Pittsburgh's Point where the Allegheny and Monongahela rivers converge to form the Ohio river,[3] serving on the governor's blue ribbon panel investigating the large crack on the just-opened I-79 bridge, and so on. I summarized what I had learned about Pittsburgh's bridges at the Second Historic Bridge Conference by sketching the trends in bridge design that started in Pittsburgh and the controversies these trends engendered.[4]

What follows is an attempt at generalizing from the above personal experiences and enthusiasms in an exploration of why historic bridges are so appealing to lay people and professionals alike.

The appeal of rural historic bridges may be tinged with an element of nostalgia. These bridges, particularly wooden ones and ones sited in remote and picturesque settings, clearly signal that they were designed for another era and not for carrying the size, weight, and volume of today's traffic. These bridges often vividly evoke the past, and the traveler sighting one of them almost expects to see a hay wagon or a steam threshing engine coming around the curve of the dirt or gravel road.

The appeal of urban historic bridges expresses itself in the opposite way. Monumental bridges, such as John Roebling's in Cincinnati and New York and Gustav Lindenthal's in Pittsburgh and New York, dominate their respective cityscapes even today. It is fascinating to try to imagine both their visual impact, when they were the largest structures in their cities, and their effect on the movement of people and goods, increasing by perhaps an order of magnitude the mobility of the people and their products. On a lesser scale, arrays of adjacent historic bridges, such as those on the Harlem River in New York, the Monongahela River in Pittsburgh and its milltown suburbs, or the Chicago River and its Ship and Sanitary Canal in Chicago, completely dominate their part of the cityscape and brutally assert their presence. On an even smaller scale, many cities, such as Columbus OH, Milwaukee WI and Providence RI, have rebuilt their historic urban bridges in order to restore the historic sense of the central cities; where this turned out not to be feasible, serious attempts were made to fashion the reconstructed bridges so as to resemble, or at least evoke, the original historic bridges.

An example illustrates this contrast in perception: the catalog of a vendor of posters on the Web offers 2800 posters in the category of "city bridges," nearly all photographs or artistic renderings of actual bridges, while the 800 posters labeled "country bridges" are mostly idyllic, pastoral views of imaginary scenes.

The restoration of historic bridges, the subject of several chapters in this book, brings on its own set of appeals. There is a great appreciation by lay people and professionals alike of the effort it takes to restore historic bridges. While the restoration of monumental historic buildings can, to some extent, be vicariously mimicked by the restoration of more modest structures by do-it-yourselfers, there is limited opportunity for applying this kind of "restoration ethic" by amateurs to bridges. Lay people, particularly builders and tinkerers, recognize and appreciate that the restoration of a historic bridge, its adaptation to new use or its re-construction at a new location all require extensive engineering effort, expertise in many long-forgotten or neglected crafts, and substantial funding, commensurate with if not exceeding the cost of tearing the bridge down and starting from scratch.

The appeal of historic bridges to professionals is a very complex subject and comprises multiple facets. I will enumerate the aspects that have appealed to me.

The first aspect that comes to mind is the sheer daring demonstrated by historic bridges. The enormous New York Central railroad bridge across the Hudson River at Poughkeepsie, the soaring Kinzua viaduct in northwestern Pennsylvania, sadly collapsed in a mild wind a few years ago, or any of the spectacular wooden viaducts on the early railroad lines in the West and many, many others are simply breathtaking.

True, many daring bridges have been built since and there are more to come, such as the proposed Messina and Gibraltar crossings, but these latter bridges are backed by solid theoretical and experiential knowledge that was not there when the early bridges were conceived and built. Similarly, a lot of daring had to be exercised to build the historic bridges with the rudimentary construction materials and technologies available at their time. More will be said about design, analysis, and construction know-how later.

Closely coupled to daring is the great resourcefulness exhibited by the designers and builders of historical bridges, frequently the same person serving both roles. Every traditional material, whether stone, brick, timber and even natural hydraulic cement, was incorporated whenever it was available and often in novel forms. Newer materials, including cast iron, wrought iron and eventually steel and reinforced concrete, were eagerly embraced as they became available and quickly put to use. Admittedly, the forms achieved with the new materials initially duplicated the forms dictated by the traditional materials and construction practices, a pattern dating back to the first iron bridge at Coalbrookdale in Shropshire, England. But very quickly resourcefulness was transferred to the innovation of structural forms and systems that took advantage of the properties of new materials. The very large number of patented bridge systems in the 19th century attests to the inventiveness and resourcefulness of the bridge builders.

A specific aspect of inventiveness was the frequent use of analogical thinking, where a previous concept is adapted to a new context to yield a novel concept. The Coalbrookdale Bridge was not based on analogical thinking: the iron arches are exact duplicates of the previously used wooden ones, including the iron pegs driven into slots in the mating iron members. In contrast, consider the Fink truss, an elaborate analog of the medieval king-post truss, the Bollman truss, an analog to suspension cables and evoking today's cable-stayed bridges with a harp configuration, and the Pauli or lenticular truss, derived by analogical thinking from the combination of arch and suspension bridges.

The oft-quoted dictum of "form follows function" is, paradoxically, both denied and supported by historical bridge forms. On the one hand, the elevations accompanying the Howe and Pratt truss patent applications reveal nothing about their respective functions: both show trusses with double diagonals. You have to go to the detailed drawings of the joints to see that the Howe joint butts the compression diagonal against the chord and threads the tension vertical through the chords, while the Pratt joint is gusseted so as to transmit the shear from the tension diagonal to the compression vertical. On the other hand, historical trusses and other bridge types are pared down to their essential members, each one exhibiting by its location, size, and shape the function it serves.

Having spent the largest part of my professional life in developing concepts and computer-based tools that assist in structural analysis and design, I am fascinated by the question of what early bridge engineers knew and when they knew it. I was educated at the time when graphic statics for truss analysis and the Williot-Mohr diagram for graphically determining truss displacements were still taught; thus I was exposed to a part of the engineers' toolkit from the latter part of the 19th century. I still find it amazing to consider how much the early historic bridge designers accomplished with their knowledge of statics and strength of materials. The 1856 textbook on bridge design used at the U.S. Military Academy at West Point covers strength of materials in 40 pages and the analysis of various structural types in 90 pages; the remaining 100 pages present detailed descriptions of specific structures. I consider it nearly inconceivable that grand arches, such as the Cabin John aqueduct near Washington DC or the Eads Bridge in St. Louis, were modeled as not much more than funicular curves and that the stupendous pre-World War I railroad bridges, such as the P&LE Bridge over the Ohio river at Beaver PA, were analyzed only with graphic statics. I find it equally amazing that all these designs were made without the aid of any kind of standard design specification, so that the designer had to rely entirely on his own mind and experience to envisage all the possible limit states that may bring the structure to collapse. Undoubtedly these early engineers learned a lot more from direct experience and observation than engineers do today; that must be the reason that they could function so well without design specifications that told them what to watch out for.

Another factor in the appeal of historic bridges has to do with the construction materials and techniques employed. These bridges become even more fascinating when one considers how they were built, that perhaps with the exception of an occasional stationary steam engine for lifting or piledriving everything was accomplished with the muscles of men and draft animals. Yet the attention to detail in stone and wood as well as iron speaks to a pride in craftsmanship that no longer characterizes the vast majority of our modern bridges.

Finally, we come to the issue of esthetics and esthetic appeal. Here two full cycles of paradigm reversals have taken place from the time of the first historical bridges in the US until today. The earliest historical bridges were absolutely utilitarian, pared down to their essential elements, with only the occasional embellishment of an ornate builder's plaque. In reaction to this stark "engineering" utilitarianism, in the latter part of the 19th century decorative elements began to appear in portals, balustrades and abutments of bridges. The turn-of the-century City Beautiful movement brought rich decorations to all elements, structural and nonstructural, of urban bridges. Structural systems themselves became highly decorated, as in the Queensborough and Manhattan bridges in New York, and in the towers of suspension bridges well into the 1930s. There was a definite synergy between the Art Deco movement, with its emphasis on expressing speed and movement, and bridges, best exemplified by the Golden Gate Bridge in San Francisco and the coastal bridges of Oregon. Then the pendulum swung back. Under the influence of architectural Modernism, a major exhibition on bridges at the New York Museum of Modern Art and the bridges of Robert Maillart and Othmar Amman, the stark, smooth, and pared-down look of bridges became the dominant esthetic for the rest of the 20th century. Today, we may be witnessing a return to a decorative phase (I emphasize the term decorative, not decorated), exemplified by the imaginative and playful forms of Santiago Calatrava's bridges. People respond positively to whichever phase of the cycles described appeals to them; one cannot argue that one phase is more esthetically pleasing than the other in any absolute sense.

I trust that this introduction provides a suitable context for the more specialized papers that follow. I invite readers to write to me about aspects of historical bridges that they find interesting and fascinating.

REFERENCES

1. White, J. and. Von Bernewitz, M. W., *The Bridges of Pittsburgh*, Cramer Printing & Publishing Co., Pittsburgh PA, 1928.
2. *Flying off the Bridge to Nowhere and Other Tales of Pittsburgh Bridges*, Documentary produced and narrated by Rick Sebak, WQED, Pittsburgh PA, 1984.
3. Tarr, J. A. and Fenves, S. J., The Greatest Bridge Never Built? *American Heritage of Invention and Technology*, Volume 5, Number 2, pages 24–29, 1989.
4. Fenves, S. J., Trends and Controversies in Bridge Design: A History of Pittsburgh's Bridges, In *Proceedings Second Historic Bridge Conference*, pp. 66–76, Ohio State University, 1988.
5. Haupt, H., General Theory of Bridge Construction, D. Appleton and Company, New York, 1856.
6. Mock, Elisabeth B., *The Architecture of Bridges*, Museum of Modern Art, New York, 1949.

Part 1

History

2 The Mississippi River Railway Crossing at Clinton, Iowa

Charles Birnstiel

CONTENTS

INTRODUCTION

Large infrastructure projects are almost always built to satisfy an economic or social need. In the case of the railroad crossing over the Mississippi River at Clinton, Iowa, the impetus was the settlement of the upper Mississippi River Valley by Europeans and its concomitant trade with the eastern United States. Because the economy of northern Illinois, southern Wisconsin, and Iowa influenced the decision to bridge the

Mississippi River at Clinton, Iowa, an initiative of the federal government to promote orderly commerce, the establishment of a Federal Mining District will be briefly described. Figure 2.1 is a map of the central Great West.

FEDERAL MINING DISTRICT

About 1720, in northeastern Illinois, southern Wisconsin, and eastern Iowa, French explorers discovered mineral deposits that the Native Americans were mining and smelting for lead. At the end of the eighteenth century Native American women and old men mined the lead and traded it to the white men who smelted it and shipped it to St. Louis, from where it was trans-shipped east. Lead was a valuable commodity for the US that was imported from England at the time because no sizable deposits of lead ores had been found in the East. Lead was, and is, used in the manufacture of paint, glass, building materials, and machinery. In order to organize the nascent mining industry a Federal Mining District was established. The federal government issued permits to miners agreeing to work claims "in a diligent and miner-like manner."[25] Opening of the mining district attracted many Irish and Cornish immigrants to southern Wisconsin and to Illinois north of the Rock River, as well as to the western bluffs of the Mississippi in Iowa. The city of Galena developed on the Fever River (now Galena River) a few miles inland of the Mississippi River and became the center of the lead trade in Illinois. Production of lead at Galena rose from 32 tons in 1825 to 27,247 tons in 1845 and then declined—because miners left Galena for adventure in the Mexican War and the California Gold Rush—to 17,082 tons in 1857 and to almost nothing in 1870.[11] Most of the Galena lead was transported to St. Louis by steamboat until 1851 when some lead was shipped east in wagons to the railhead of the Galena and Chicago Union Railroad (G&CU), then at Rockford, IL. After the G&CU reached Galena in 1857 the portion of the lead sent east to Chicago exceeded that shipped to St. Louis.

The trend for shipping lead eastward from Galena via rail also applied to other commodities, such as grain. Chicago was to become a major grain market due in large measure to the perfection of steam-powered grain elevators by the G&CU and other railroads, and to the development of a system for grading, warehousing, and trading grain. The effect of these developments drew grain trade to Chicago that had formerly been shipped by steamboat to St. Louis and New Orleans. As early as 1851, despite the lack of a Mississippi River bridge, an Iowa merchant remarked "it is estimated that four thousand teams cross the river from the Iowa side to Galena and return in a year."[11] These passages were made using ferries in the summer and over the river ice in the winter. Besides transporting agricultural products from the west to the east side, there was also the need to transport manufactured goods and immigrants in the reverse direction. Chicago was becoming a center for the farm machinery and clothing industries.

To gain an appreciation of the rapid increase in population in the valley during the middle of the nineteenth century, consider population changes in Illinois and Iowa. Illinois became a state in 1819 despite the fact that its population was less than the minimum of 60,000 stipulated by Congress some 30 years earlier. It now has an area of 56,345 square miles. The population was 55,000 in 1820, 476,000 in 1840,

FIGURE 2.1 Central part of the Great West.

and 1,711,000 in 1860,[1] a 30-fold increase in 40 years. Iowa was admitted to statehood in 1846 with an area of 56,275 square miles. The population was 43,000 in 1840 and 675,000 in 1860, a 15-fold increase in 20 years.[1] Based on these figures, the population in both states doubled every two years and this was reflected in the commercial environment. It was obvious that a railroad bridge across the Mississippi River between Iowa and Illinois would benefit the Galena and Chicago Union Railroad and the economy of Iowa and the western lands. The route the railroad took toward that goal is briefly described below.

GALENA AND CHICAGO UNION RAILROAD

Railroads came late to Chicago. Steam locomotives had been puffing in the East for two decades before the first run of the "Pioneer" on November 20, 1848 on Chicago's first railroad, the Galena and Chicago Union Rail Road, "the only one ever to be controlled and backed by city entrepreneurs."[16] The late start of railroad construction in Chicago was mainly due to the attitude of eastern capitalists (mostly in New York) who viewed Chicago as a Great Lakes port, which it rapidly and successfully became after the city was incorporated in 1833. An astounding number of vessels would dock in Chicago daily during the summer. Easterners had invested heavily in water transportation systems that included shipping on the Great Lakes, facilities on the Erie Canal and terminals in New York harbor. They feared that railroads would detract from that commerce. However, the developing Chicago industry needed year-round transportation. In January 1836 the Illinois legislature passed an act incorporating the Galena & Chicago Union Rail Road Company.[19] A stretch of four miles from Chicago was built in the summer of 1838, except for the running rails.[6] Construction was not resumed in 1839. From the first G&CU annual report, it appears that the work was abandoned.

William Butler Ogden and friends purchased the G&CU stock in 1847 with the intention of completing the road as originally planned. However, they could not obtain eastern capital for further construction and instead sold stock to the residents of towns along the proposed route. A corps of engineers was employed and the first division, from Chicago to Elgin, was completed at the end of 1848. It was a single track built with strap rail. According to the eighth G&CU annual report of June 1855, the second division, from Elgin to Rockford was put into operation in August 1852; the third division, from Rockford to Freeport, was completed in September 1853. By the summer of 1855 all the strap rail was replaced with wrought iron T-rail weighing 56 pounds per yard, and a stretch near Chicago had been double-tracked. Construction had started on a route from Chicago to Fulton, IL, and operated as far as Dixon in December 1854. The stretch between Dixon and Fulton was leased from the Mississippi and Rock River Junction (M&RRJ). The line was open to Fulton in December 1855.

CHICAGO, IOWA AND NEBRASKA RAILROAD

The Chicago, Iowa and Nebraska Railroad (CI&N) was organized by the Iowa Land Company, a development group formed in 1855. They replatted an earlier town and named it Clinton in honor of DeWitt Clinton, governor of New York State at the time the Erie Canal was built.[17] Track was laid southward to Camanche in 1857 and continued westward to the town of Vandenburg, which they renamed DeWitt, and reached Cedar Rapids, about 82 miles from Clinton, in 1859.

When the G&CU reached Fulton, IL, on the east bank of the Mississippi in 1855, the residents of Lyons, on the west bank just opposite Fulton, assumed that the Mississippi River crossing would be anchored in their town, especially as it was larger than either Clinton or Camanche. The Lyon and Iowa Central (L&IC) planned to cross the Mississippi River on a high-level fixed bridge that would have avoided the need for a swing span, as with the present low-level bridge. The principal L&IC promoter,

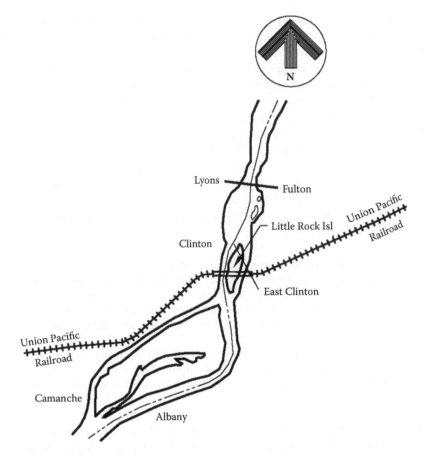

FIGURE 2.2 Vicinity of Clinton, Iowa.

H. P. Adams, sold a considerable amount of bonds to the locals, but after only eight miles were graded, he absconded with the funds and "left the counties with taxes to pay and Irish laborers to feed."[7] When the CI&N changed the proposed alignment of the Mississippi River bridge from Lyon–Fulton to Clinton–Fulton, the town of Clinton grew and eventually absorbed the smaller neighboring towns of Lyons, Chancy, and Ringwood. Figure 2.2 shows the location of these towns and the two bridge alignments. The first and the present swing bridges are on the same alignment.

PLANNING THE MISSISSIPPI CROSSING

Evidently, the G&CU started planning for a Mississippi River bridge as early as 1857, engaging William J. McAlpine with the title assistant to the president and chief engineer. For the first two decades of his civil engineering career, McAlpine worked on New York canals and waterworks. From 1850 to 1857 he was New York railroad commissioner, then became chief engineer and assistant to the president of the Erie Railroad. He joined the G&CU with the same titles in 1857. The fourteenth annual G&CU report does not list him in that position. W. J. McAlpine had a prominent engineering career as a manager, was the third president of the American Society of Civil Engineers, and was the first American elected to the Institution of Civil Engineers of Great Britain.

McAlpine's younger brother, Charles Legrand McAlpine, was "of more retiring disposition" and held similar engineering positions in New York State. Later he oversaw construction for various railroads through a consulting office in New York City. It was C. L. McAlpine who submitted a report, dated June 23, 1858, to the G&CU for a proposed single-track bridge across the Mississippi River that included preliminary cost estimates for five different alignments.[13]

In October 1858, the G&CU directors proposed leasing a railroad bridge from Fulton, IL, to Lyons, IA, that was to be built by an independent company strictly controlled by the G&CU. The CI&N, however, made known its intention to bridge the Mississippi—to the considerable displeasure of the G&CU. In February 1859, these directors reported,

> They [CI&N] announced their determination to construct a bridge over the East chan-
> nel of the Mississippi river, two miles below Fulton, to Little Rock Island, and from
> thence to connect with Clinton, opposite, for the present, by a ferry, but with a view of
> ultimately constructing a drawbridge over the Western, or main channel. They propose
> to connect their tract with our road [G&CU], either at Fulton or about three miles east
> of that point.
>
> The movement on the part of that Company [CI&N] the Directors [G&CU] do
> not approve ... but hope to effect such an adjustment of the matter as will result in
> the construction of the bridge at Fulton, under direction and control of the Company
> [G&CU].

The G&CU intended to take strong measures to control the bridge construction.

Meanwhile, the CI&N was carrying out its own intentions. A year before, on February 14, 1857, the Albany Bridge Company (also, Albany Railroad Bridge Co.) was incorporated in Illinois to build a railroad bridge across the Mississippi.[22] Among the original stockholders was William W. Durant, the notorious railroad stock manipulator. The CI&N began constructing the bridge between a point two miles south of Fulton on the Illinois side to Little Rock Island, completing that portion of the bridge before January 1860. Railroad cars were ferried across the main channel (from the island to Clinton) on CI&N steamboats, thereby opening a railroad route from Chicago to Cedar Rapids, Iowa. The G&CU leased the CI&N, and the Albany Bridge Company was leased to the CI&N. Thus, by August 1, 1862, the G&CU controlled the Chicago to Cedar Rapids line. It was consolidated into the Chicago and North West Railway on December 10, 1869 (from partial report submitted to the Interstate Commerce Commission[19]).

THE FIRST BRIDGE

GENERAL WARREN'S REPORT

As construction drawings and specifications for the first railroad bridge over the Mississippi River at Clinton have not surfaced, recourse will be to the description in the 1878 report of Brevet Major-General G. K. Warren for the US Army Corps of Engineers. Warren first made observations and surveys at the bridge site in 1866–67, in connection with his research on bridging the river and other navigational matters between St. Paul, MN, and St. Louis, MO. He was assisted at Clinton by three civil engineers: J. P. Cotton; W. Weston; and A. M. Scott. Despite Illinois's authorization of the Albany Bridge Company to construct a Mississippi River bridge whose eastern segment to Little Rock Island was completed before 1860, Iowa did not authorize construction over its territory until three years later. The bridge opened on January 1, 1865 and it had been built without federal authority. It was legalized *post facto* by Congress on February 27, 1867. The information below is based mainly on Warren and emphasizes the swing span. Figure 2.3(A) depicts the elevation of the first bridge when looking upstream. The vertical scale of the diagram is many times the horizontal scale. It is important to recognize that low first cost was the primary consideration for western railroad construction. Nowadays, engineers design for a structural life of between 80 and 120 years with safety as the primary consideration. It was not always so! A design life of ten years for railroad construction, except for passenger stations and certain non-track facilities, was considered adequate while railroad technology was developing so rapidly. Also, except for wrought-iron rails (which were at first imported from England) and machinery, most construction material was obtained locally. Like an old-time army, the western railroad "lived off the land." Railroad developers realized that continual upgrading of track and bridges would be necessary. As no other contemporary descriptions of the first bridge were found, related excerpts are from Warren (1878). The material will be presented under separate headings for the Illinois bridge section (east bank of Mississippi to Little Rock Island) and the Iowa bridge section (Little Rock Island to west bank at Clinton).

Illinois Bridge Section

Commencing on the left or Illinois shore, there is 1,300 feet of wooden trestling in the approach to the bank of the river; then there were at first 7 spans of wooden truss, each 200 feet long; then a causeway 583 feet long across Little Rock Island to the Iowa Channel … The span next to the Illinois shore … at the time of our survey, in 1866, this bridge was in very bad condition, the piers, built with a core of concrete, were cracked in several places, and to support the spans temporarily various devices had been resorted to. The trusses, which were of the pattern known as McCallum's, were decayed in many parts, and a large number of the diagonals were broomed where they abutted against the angle blocks. [See Figure 2.3.]

In 1868 and 1869 this part of the bridge was entirely rebuilt; the first span next to the Illinois shore was replaced by an iron truss; the remaining portion, 1,200 feet to the abutment on Little Rock Island, was replaced by 8 iron spans of 150 feet each; the new piers were smaller, 9 feet wide at the top, of masonry, resting upon piles. The 200-foot span and the two adjacent 150-foot spans were built by the American Bridge Company, of Chicago, on the Post pattern. The next two spans were built by the Detroit Bridge Company, Detroit, Mich.; the next two, by the Phoenixville Bridge Company, of Phoenixville, Pa.; the next two, by the Keystone Bridge Company, of Pittsburgh, Pa.; the last six are of the Pratt truss pattern.

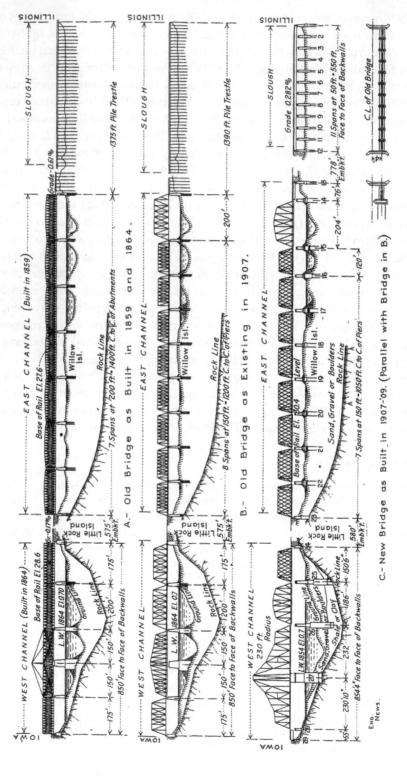

FIGURE 2.3   Profiles of the three C&NW bridges over the Mississippi River. (From Bainbridge.[2])

Iowa Bridge Section

The bridge over the Iowa channel, built in 1864–65, consists (commencing at Little Rock Island) of one span 174 feet and 200 feet of wood; then a pivot draw-span of iron (Bollman) 300 feet, over all, with an opening on the east side of pivot-pier of 119 feet at low water, and one of 128 feet on the west side; then a span of 174 feet to abutment on the Iowa shore (this of wood); then a short piece of embankment.

SWING BRIDGE TYPOLOGY

Swing bridges are categorized by the type of bearing that supports the draw when it is in an open position.[9,20] If the whole self-weight is supported by the pivot the span is termed center bearing. The weight is usually balanced on the pivot bearing. To prevent the draw from tipping under unbalanced loads, such as wind, from four to eight balance wheels are provided that roll on a large diameter circular track concentric with the pivot. When the draw is balanced those wheels normally clear the track by about 0.2 inches. Under a sufficiently strong lateral wind the draw tips and some balance wheels contact the track and the resultant reaction equilibrates the overturning moments. Movable wedges driven under the trusses support the live load when the draw is fully closed.

Swing bridges whose dead load is supported by a large-diameter circle of tapered (conical) rollers when in an open position are termed rim bearing. The rollers run on a circular track whose diameter is approximately the same as the spacing of the outer superstructure trusses or girders. When the bridge is closed the rim bearing supports both dead and live load.

When both the pivot and rim bearings support dead and live loads, the swing bridge is termed combined bearing. The type was controversial because distribution of load between the pivot and rim bearings was considered ambiguous by some engineers. However, a century ago, some engineers used the combined bearing type for large and heavy bridges. Some structural framing schemes were developed which assured reasonably reliable distribution of load between pivot and rim bearings. The topic is beyond the scope of this paper except that the present Clinton bridge is a combined type. Controlling the load distribution by jacking was attempted, as is discussed in a later section of this paper.

THE FIRST SWING SPAN

The first swing span, completed in 1864, was a pair of Bollman patent suspension trusses with inclined stays (hog-ties) radiating from the top of a cast iron tower mounted on the pivot bearing assembly, as depicted in Figures 2.4 and 2.5. The bridge may have been of the rim bearing or combined bearing type.

When the span was in the open position, the trusses cantilevered from the pivot assembly and were supported by tension in the inclined rod stays. In the closed position, the trusses were simply-supported, i.e., one end of each truss rested on the pivot assembly and the other on an adjacent pier. A steam engine operated gearing to rotate the span.

Wendell Bollman patented the first all-metal American railroad truss.[3] It had cast iron compression members and wrought-iron tension members and was designed so that local members could be replaced without shoring the whole span. Figure 2.6 is

FIGURE 2.4   First swing span in open position. (Courtesy of C&NW Historical Society.)

FIGURE 2.5   First swing span in closed position. (Courtesy of Clinton County Historical Society.)

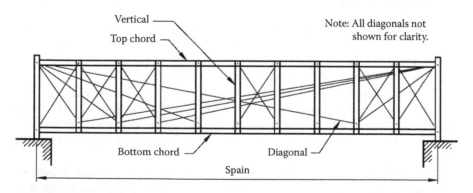

FIGURE 2.6   Bollman patent suspension truss bridge.

a simplified diagram of his invention. The Baltimore and Ohio Railroad replaced wooden truss bridges that required extensive repair with Bollman trusses. The first railroad bridges had wooden chords and web members that deteriorated with time, especially under increasing locomotive weights. As was standard practice at the time, many contractors used locally sawn timber when it was green, and this lead to decay and shrinkage after the members were built into the bridge, resulting in deformations that loosened truss joints and increased dead load deflections. This was compensated for, to a degree in combination bridges, by periodically retightening rod diagonals. The engineers' reports included in the G&CU annual reports frequently mention replacing chords of timber trusses and whole bridges. In 1862 a new bridge was brought into use across the Fox River near Elgin, IL, consisting of three spans of 125 feet each. "The bridge is of iron, of the style known as the patent of Wendell Bollman, of Baltimore." Thus, the G&CU had experience with the Bollman trusses before they were selected for the swing span of the bridge at Clinton.

The writer could not ascertain the names of the fabricator and erectors of the Clinton draw. Darnell[5] indicates that W. Bollman and Company may not have been operating from 1861 to 1863 and that his Patapsco Bridge and Iron Works was organized about 1865. It would have been improbable for Bollman to have been directly involved in building such a large bridge under those circumstances. It is also of interest that a cable-stayed swing bridge based on Bollman's patent trusses and with a 200-foot draw was built in 1868 across Quincy Bay (a channel of the Mississippi River) and that it existed in 1899.[12]

The Clinton bridge was the second railroad bridge built across the Mississippi River below St. Paul, MN. It was also the second of the first generation of Mississippi River bridges, those constructed before about 1880. The draws of these were mostly proprietary designs of various bridge building companies such as the American Bridge, Baltimore Bridge, Detroit Bridge and Iron, Keystone Bridge, and Kellog Bridge companies.

REPAIRS TO FIRST BRIDGE 1873 TO 1887

Repairs not chronicled in Warren[24] included replacement, in 1874, of the timber Howe trusses over the west channel by wrought-iron Whipple trusses that were erected by the American Bridge Company.[22] The Whipple trusses appear in Figure 2.3(A), flanking the Bollman swing span. The first span over the east channel (200 feet) was replaced by a Whipple truss span in 1880. In 1885, Pratt trusses replaced the remaining eight 150-foot truss spans over the east channel. This litany of truss span replacements should not be considered too unusual because it occurred during a period when the weight of rolling stock increased dramatically. Railroads also procured major bridges on a design-build basis. In this process, the contractor usually offered a fixed price for a turn-key completion of the project to the owner. The problem was that competing contractors did not design to the same criteria. Most railroads did not have engineering staffs able to specify structural bridge design requirements and were even less able to monitor and enforce them. They generally left the structural design of the superstructure to the contractor, and, as the lowest bidder usually received the purchase order, the railroad got what it paid for at best.

No generally accepted industry-wide design and material specifications existed that railroads could reference in their dealings with contractors. The American Society for Testing and Materials (ASTM) was not organized until 1898, and the American Railway Engineering Association (AREA) issued its first *Manual of Recommended Practice* in 1905. Charles Schneider proposed the first movable bridge design specification in 1908,[21] but it was not adopted by AREA until 1922. Not until railroads engaged independent bridge consulting engineers, such as Theodore Cooper, George S. Morison, and J. A. L. Waddell, who had developed their own in-house design specifications and procedures, did the quality of bridgework improve.[18,23]

THE SECOND BRIDGE

In 1887, after twenty years of service, during which locomotive weights had increased considerably, the Bollman swing span was considered inadequate and was replaced with a Pratt truss draw built by the Detroit Bridge and Iron Works.[12] The new draw trusses were single-intersection type with counters (members capable of resisting tension only, used where the panel shear changes sign on passage of a moving load) as in Figure 2.7. In 1898, the 200-foot Whipple truss over the east channel and the three fixed spans over the east channel were replaced by Pratt trusses also built by the Detroit firm. Figure 2.3(B) shows the whole bridge at this stage. Part of the Pratt draw is depicted in Figure 2.8, photographed from the adjacent high-level wagon bridge built in 1891. It was still a single-track bridge designed for Coopers E-40 loading, the common loading prior to 1900.[26] No design or construction drawings or specifications for the Pratt draw, nor the name of a consulting engineer, have been found. The type of bridge was typical for the second generation of bridges across the Mississippi. The upper and lower treads of the rim bearing and the lifts at the ends of the draw were replaced in 1901, only 14 years after they were installed.[2] The draw span was taken out in 1909 and scrapped.

THE THIRD BRIDGE

The trend toward heavier rolling stock and as many as 150 train movements per day over the single track, caused the Chicago and North Western to consider replacing the whole bridge with a double track bridge on the alignment of the first bridge as early as 1900.[2] At that time many slow-orders were issued at the bridge and this contributed to traffic congestion. A slow-order is a direction (usually temporary) to locomotive engineers that train speed shall be limited to a given value over a specified length of track. They are issued where track geometry does not meet the standard for the normal speed on a stretch. Speeds higher than the slow-order might cause derailment of a locomotive or a car, a matter of considerable expense to the railroad. Track geometry changes sometimes occur at bridge abutments due to climatic changes and anything that changes the supporting conditions, such as soil settlement or flexibility of a bridge floor system. The joints between the ends of the draw and adjacent fixed spans are especially vulnerable. Surveys and borings were made for the new double track structure in 1901. However, before design and construction could start, it was

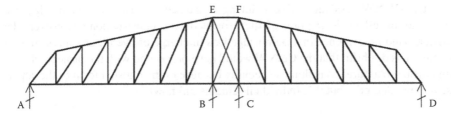

FIGURE 2.7   Second swing span truss.

FIGURE 2.8   Second swing span in closed position. (Courtesy of Clinton County Historical Society.)

necessary to decide on the type of swing bridge and the shape of the superstructure trusses. So, before describing the existing bridge we shall consider the engineering organization and some factors that had to be considered in making the choice of bridge type and form.

ENGINEERING ORGANIZATION

The writer was unable to ascertain the name of the consulting engineer, if any, for this project. Many of the engineering design drawings for both the superstructure and the machinery and the shop drawings[4] have been examined and no notation regarding a consulting engineer was found. The title blocks on the design drawings read C&NW Ry, East Iowa Division, Bridge No. 0¾ over Mississippi River at Clinton, Iowa, Office of the Engineer of Bridges, Chicago, IL. Some are signed by S. J. Stern, engineer of bridges and W. H. Finley, asst. chief engineer. None have an approval signature of the chief engineer. The shop drawings for the superstructure steel and the mechanical machinery were prepared by the Pennsylvania Steel Company of Steelton, PA, a technically progressive steel manufacturer that was absorbed into the Bethlehem Steel Company in 1916.

The C&NW resident engineer during construction was F. H. Bainbridge. He wrote an excellent description of the project during construction.[2] However, the writer found no evidence that Bainbridge was in charge of overall or superstructure design. It is doubtful that the C&NW would have proceeded with the project without an outside structural consultant, especially because of the unusual form of the superstructure selected—but this individual remains unknown.

CONTEMPORARY SWING BRIDGE FORMS

During the period 1890 to 1910 at least 400 swing bridges were constructed in the United States. Their structural forms varied mainly to obtain required stiffness and strength; and the effort required to make numerical structural analyzes of externally and internally indeterminate structures. For the case of rim bearing and combined bearing swing bridges, there was the additional consideration that live load should not cause uplift at the girder/truss reactions on the rim bearing drum girder. For example, let us consider the rim bearing swing span truss of Figure 2.7. According to the conventional beam theory (bending strains only), there would be uplift at support B due to a live load placed on span CD. Because span CD is so much longer than BC the uplift at B could, theoretically, be quite large. Of course, the uplift at B would be counteracted by downward dead load. However, for open-deck railroad bridges, the ratio of live to dead load is large and, according to conventional beam theory, the net reaction could be an uplift force which would be impractical to accommodate in the rim bearing. Fortunately, trusses were usually used as spanning members, instead of solid web girders, and their shear deformations are more significant than shear deformations in solid web girders with the result that the uplift at B for a truss would be much smaller than for a beam. Nevertheless, designers wanted to minimize the possibility of uplift at B, so the truss diagonals in panel BC were omitted or made so flexible that they were ineffective in transmitting vertical shear across the panel. Configuring the truss so that Panel BC could not transmit shear also had the advantage that it reduced the degree of indeterminacy by one. This was important because it significantly reduced the complexity of structural analysis computations.

While the whole topic is beyond the scope and available space of this paper, let us consider a simple matter: external determinacy of the planar beam in Figure 2.9(a). If the beam has supports that can develop the reactions R_1 through R_3, the system is stable and the values of the reactions can be computed from the three available equations of statics; the sum of horizontal and vertical forces each equals zero and the moment of all forces about an axis normal to the plane of the structure equals zero. (Moment is the product of force times distance). The computation is independent of the deformational response of the beam to the forces (for first order analysis). Because we have three equations of statics available we can solve for the three unknown reactions without difficulty.

Now suppose the beam is supported by an additional reaction R_4 as in Figure 2.9(b). There are four reaction components but we have only three equations of statics available, insufficient for computing the four reactions. The system is said to be externally indeterminate to the first degree because we have one more unknown

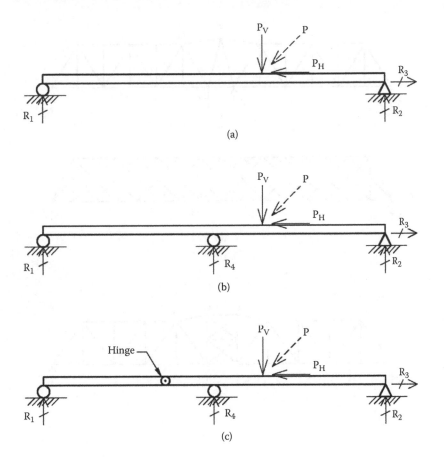

(a)

(b)

(c)

FIGURE 2.9   Beam diagrams.

reaction than available static equations. An additional relation is necessary. It may be obtained by considering the elastic properties of the beam as described in any structural analysis textbook, such as Hibbeler,[8] Kassimali,[10] or Michalos.[15] Solving this problem is not difficult with modern techniques but required effort in 1890. One way of avoiding the difficulty is to modify the structure as in Figure 2.9(c) to make it determinate by inserting a physical hinge at some point on the beam. This hinge (assumed frictionless) is known as an equation of condition. It specifies that the bending moment is zero, i.e., the beam has no bending rigidity at this point. So now we have four equations, the three equations of statics plus the equation of condition, with which to solve for the four reactions. (However, the presence of the hinge changes the behavior of the beam under loading.) Trusses can be treated in a similar fashion; but, instead of adding a hinge, a bar is removed. The application will be illustrated by considering some forms of swing bridge trusses built in the period 1890–1910.

Figures 2.7, 2.10 and 2.11 show some popular forms of swing span trusses used for rim bearing swing bridges in the bridge-closed position. In each diagram A and D are the ends of trusses at which machinery is located to lift or lower the trusses. B and

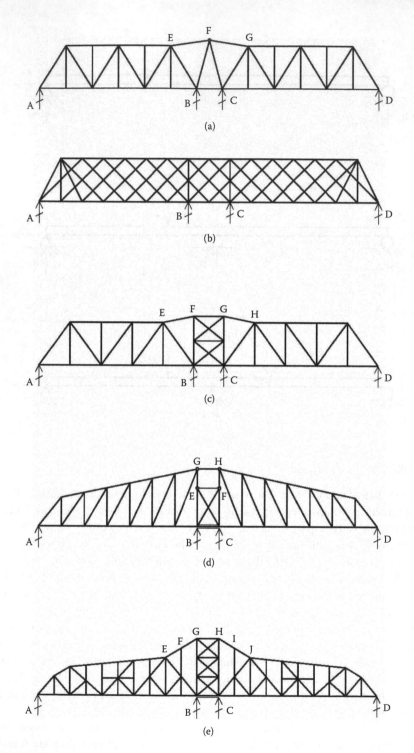

FIGURE 2.10 Some swing span trusses built or proposed 1890–1920.

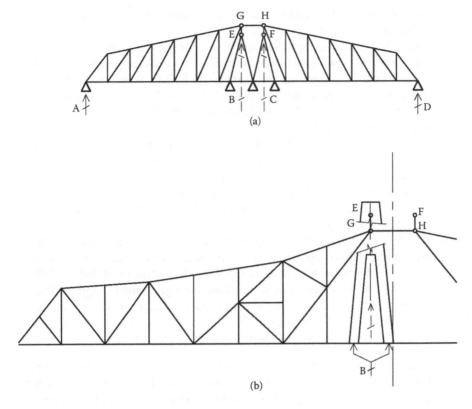

FIGURE 2.11 Swing span trusses supported by A-frames.

C are supports for the trusses at the rim bearing. For each truss, the features of equations of condition and the structural analysis for the case of gravity loading will be described. Because it is stipulated that there are no horizontal loads, the static requirement that the sum of the horizontal forces equal zero is, by definition, satisfied.

Figure 2.7. Pratt truss with weak diagonals BF and CE. With BF and CE inactive no vertical shear can be transmitted across the center panel. This is an equation of condition. Reactions = 4. Available eqs. = 2 + 1 = 3. Indeterminate to first degree.

Figure 2.10(a). Warren truss with verticals. Bars EF and FG are eyebars which can only resist tension. When ends A and D are lifted these bars buckle and are effectively removed from the system. Reactions = 4. Available eqs. = 2 + 2 = 4. Determinate. No live load shear transfer across Panel BC because of buckled bars.

Figure 2.10(b). Lattice truss. Reactions = 4. No equations of condition. Available eqs. = 2. Indeterminate to second degree. Used for many English swing spans. C&NW built such a bridge over Kinnickinnic River in 1899. Uplift of truss at B and C possible, depending on live load.

Figure 2.10(c). Modified Warren truss. Bars EF and GH are tension-only eyebars, giving two equations of condition. Reactions = 4. Available eqs. = 2 + 2 = 4.

Determinate. No live load vertical shear transfer across Panel BC because of buckled top chord.

Figure 2.10(d). Pratt truss. Bars EG and FH are compression links that support the truss. The center panel BEFC is rigid and acts as part of the pivot pier. No shear can be transmitted across the panel E, F, G, H and this is an equation of condition. Reactions = 4. Available eqs. = 2 + 1 = 3. Indeterminate to first degree. No live load shear transfer across BC because of lack of diagonals EH and FG.

Figure 2.10(e). Phoenix truss. Bars EF, FG, HI and IJ are eyebars. As ends A and D are lifted, these may buckle. Hence EG and HJ are equations of condition. Reactions = 4. Available eqs. = 2 + 2 = 4. Determinate. Combined bearing swing span 520 feet long built over Williamette River in 1907. Operated to 1989. No live load shear across Panel BC because of buckled bars in top chord.

Figure 2.11(a). Pratt truss. Effectively same as Figure 2.10(d). Quadrangular rigid frame of 10(d) replaced by two A-frames. Indeterminate to first degree. No live load shear transfer across BC because lack of diagonals EH and FG.

Figure 2.11(b). Half elevation of existing Clinton bridge. Same as Figure 2.11(a) except that A-frames are raised above the top chord of the truss so that links are in tension. Indeterminate to first degree. No live load shear transfer across Panel BC because diagonals EH and FG are lacking.

The above seven examples of approaches for simplifying swing bridge trusses by considering them partially continuous or simply supported in the closed position are not necessary today. The computer now makes possible the analysis of almost any elastic structure using the displacement method in matrix form.[14]

CONSTRUCTION OF THE THIRD BRIDGE

Congress authorized reconstruction of the bridge, and the secretary of war approved plans in May of 1907. The contract for the entire substructure was placed with the Foundation Company (New York) less than two weeks later. Structural steel for the west channel spans was let to the Pennsylvania Steel Co. of Steelton, PA, and for the structure east of Little Rock Island, to the Wisconsin Bridge and Iron Co. of Milwaukee, WI. Figure 2.3(C) shows the profile of the new bridge and its relation to the second bridge. A larger scale profile of the structure over the west channel is shown in Figure 2.12. A photograph of the closed bridge, taken in October 2007 from the southwest, appears as Figure 2.13. Every component visible in the photograph is thought to be original—at least 98 years old. A view of the fully open draw, taken from the end of the east flanking span (looking west), is in Figure 2.14.

The details of the foundation construction, as presented by Bainbridge, are very interesting but space restrictions preclude a thorough review. As is evident from Figure 2.3(C), three piers for the west channel bridge are quite deep and the pivot pier is founded on rock. Pier Nos. 25, 26 and 27 were built using the pneumatic process. Timber caissons were constructed on ways on the west bank, launched into the river, and towed into position. The caissons had timber decks six feet above their bottom edges to form a working space for "groundhogs." The soil was excavated under air

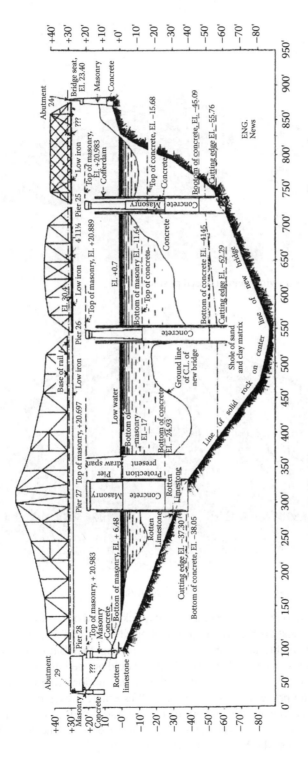

FIGURE 2.12 Profile of third bridge over the West Channel. (From Bainbridge.[2])

FIGURE 2.13   Third swing span in closed position.

FIGURE 2.14   Third swing span in fully-open position.

pressure sufficient to prevent leaks that could quickly fill the chamber and drown the workers. The greatest air pressure required at pier No. 26 was 34 pounds per square inch, more than double atmospheric. It is interesting that Bainbridge calls them "the usual type of wooden deck caisson." They would not be built today because the enormous amounts of timber required for their construction are no longer available and the pneumatic process is seldom used by Western countries for safety reasons. The Foundation Company (a New York firm) built many bridge piers in the Midwest using the pneumatic process. As late as 1929, they used it for the river piers of the highway bridge across the Missouri River at Hermann, MO.

Swing Span Superstructure

The moving live load assumed for each track was Cooper's E-50, and the structural design generally followed the specification developed by C.C. Schneider for the American Bridge Company. The trusses were spaced wider apart than was usual at the time so that derailed car wheels would bear against the track guard timbers before the car bodies struck the trusses. The length of the draw is 460 from feet center to center of the end floorbeams. The draw is a combined bearing type in which superstructure load is distributed to both the rim bearing and the pivot bearings. The rim bearing comprises a nest of tapered rollers running between tread plates, the lower one mounted on the pivot pier and the upper one fastened to distribution framing on the underside of the draw superstructure. The distribution framing spreads the vertical dead and live load circumferentially so that the rollers are approximately equally loaded. Usually the distribution member is a circular drum girder; or, for very heavy bridges, two concentric drum girders. For this bridge, girders arranged in the form of an octagon in plan perform the function of a drum girder. Figure 2.15 is a transverse cross-section which shows the pivot bearing and the rim bearing. Figure 2.16 shows the rim bearing components: the tapered rollers, the live ring, the upper and lower treads, and the castings on which the treads are mounted. All the visible components are original.

One of the difficulties in designing combined bearing draws is the question of load distribution between the pivot and rim bearings. The designer chose a configuration that the writer believes to be unique. Two pairs of A-frames within each truss (four total for the draw) distribute the end shears of each half of the truss to a square panel of girders in the floor system that has additional diagonal girders arranged to form the octagon drum. Figure 2.17 is a photograph taken from the end of the east flanking span with the draw fully open, showing the pair of A-frames in the south truss and the girder on which they bear. Note the tension links at the tops of the A-frames from which the truss is suspended. In order to control the distribution of load between the rim and the pivot bearings, the trusses were jacked upward and connections made in a manner such that the rim bearing and the pivot bearing were, initially, loaded almost equally.

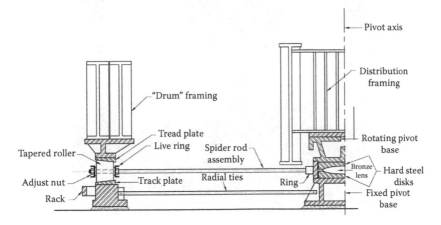

FIGURE 2.15 Transverse cross-section through pivot.

FIGURE 2.16   Rim bearing assembly and span drive pinion.

FIGURE 2.17   A-frames viewed from flanking span.

FIGURE 2.18   Primary span drive in Operator's House.

TURNING MACHINERY

The draw is rotated by an electro-mechanical drive. As of October 2007, all the mechanical components were original, except for pinion shafts that had been replaced because cracks had developed at stress risers at shaft keyways, not an uncommon occurrence. The electrical system has been modified over the years. Originally, a dedicated powerhouse on the Clinton shore contained two 50-HP Otto-cycle gasoline engines to drive DC generators and a large bank of lead-acid cells. Power was delivered to the span drive, located in the operator's house, through a submarine cable. The draw could be operated directly from the generators, or from the battery alone, or some combination of the two. Now power is supplied from a utility via overhead cables, and there is an emergency engine-generator with a transfer switch.

The mechanical transmission system is simple, but of an unusual arrangement. It converts the high-speed, low-torque output of the electric motor to slow-speed, high-torque for the final gear sets that rotate the draw. Four pinion shafts engage the common rack. Equalizers (differentials) in the transmission system assure that the pinions are equally loaded, or nearly so. Figure 2.18 shows the primary drive in the operator's house. The motors are original, but have been refurbished. Figure 2.19 depicts some of the open intermediate reduction gearing that is mounted within the truss. This is very unusual but is efficient in terms of space and accessibility.

END MACHINERY

The lifts at each end of the draw raise the deflected ends of the cantilevered trusses so that rails on the fixed and movable structure are on the same level when the bridge is closed and ready for rail traffic. They are cam operated from a gear train powered

FIGURE 2.19   Span drive intermediate gearing mounted within truss.

by an electric motor, originally 20 HP but since increased to 40 HP. Except for the
motor, the system is original, including the limit switches. The centering device that
locks the rotating bridge in the closed position is a horizontal bar that is driven by an
air cylinder for fast action. It, too, is original.

SUMMARY

Political and economic conditions in the Midwest strongly influenced the development of the Galena and Chicago Union Rail Road and the construction of a railroad
bridge across the Mississippi River at Clinton, Iowa. European immigration to the
area through New Orleans in the early 1800's, and later via Great Lakes navigation,
created the demand for railroad service. Capital from New York financed the building
of Chicago and the internal improvements in the surrounding land. Because this work
was undertaken at the beginning of industrialization in the United States, mistakes
were made, but they led to the development of civil engineering as a discipline and
educational institutions for transmitting that knowledge.

By comparing the three movable bridges constructed at Clinton, the present one
being 98 years old, one gains an appreciation of how fast structural engineering and
steel construction methods progressed in the century after 1850. The three biggest
were vastly different in design and construction because of technological changes in
railroad equipment, ferrous metallurgy and manufacturing processes, construction
methods, and engineering made during the 45 years between completion of the first
and the present bridges.

The 1865 bridge comprised two standard fixed bridges hung end to end by rods
from a cast iron tower mounted on a turntable atop the pivot pier. The draw was
300 feet long. It was the first wrought-iron swing bridge across the Mississippi below
St. Paul and the first operated by steam. The second draw, opened in 1887, was also
single track, built on the piers of the first swing span. The superstructure consisted

of wrought-iron Pratt trusses with counters designed by an unknown engineer. The present bridge, the third, was opened in 1910. It is double-tracked and much more substantial than the prior bridges, because, in part, the draw is 460 feet long between end floorbeam centers. Although design and shop drawings were made available by the present owner, the Union Pacific Railroad, the name of the consulting engineer, if any, could not be ascertained. The special feature of this draw is that the trusses are hung from A-frames mounted on the turntable. No other swing span of such design is known to the writer. The Chicago and North Western swing bridge at Clinton is in good condition, despite the severe operating conditions over many decades. It has unique engineering features of historical importance.

ACKNOWLEDGEMENTS

Many persons provided the author with information or other assistance in preparing this paper and their help is acknowledged with thanks. Among them were; Brad Crouch with the Clinton County Historical Society, Kim Limond and Fran Buelow with the Clinton Public Library, the Mercantile Library of St. Louis, MO, University of Dubuque Library, Dr. Alfred Korn, the late Alexander H. McPhee, and David A Simmons. Special thanks are due to associates of the Union Pacific Railroad, listed alphabetically; T. J. Ballard, Douglas Batey, Michael P. Freeman, Linda Hogan, Chris T. Keckeisen, Mark L. McCune, M. Kevin Moran, and Mike Ray.

REFERENCES

1. Americana, *The Encyclopedia Americana*, International Edition, Grolier, Inc., Danbury, CT, 1991.
2. Bainbridge, F. H., "New Bridge Crossing the Mississippi River at Clinton, IA; Chicago & Northwestern Ry.," *Engineering News*, January 21, 1909, pp. 63–69.
3. Bollman, Wendell, US Patent No. 8624 for "Suspension Bridge", Jan. 6, 1852.
4. Chicago & North Western Railroad, *Selected Drawings and Documents,* Union Pacific Railroad [earlier Railway], Omaha, Nebraska, 1907–1909, retrieved in 2007.
5. Darrell, Victor C., *Directory of American Bridge Building Companies*, Occasional Publication No. 4, Society for Industrial Archeology, Washington, D.C., 1984.
6. Gamst, Frederick C., editor, *Early American Railroads: Franz Anton Ritter von Gerstner's 'Die innern Communicationen 1842–1843'*, Translated by David J. Diephouse and John C. Decker, Stanford University Press, Stanford, CA, 1997.
7. Hemminger, Tom and Guhr, Robert C., "A Historical Sketch of Clinton, Iowa," *North Western Lines*, Elmhurst, IL, Spring 1993, pp. 48–79.
8. Hibbeler, Russell C. *Structural Analysis*, 5th Edition, Prentice-Hall, Upper Saddle River, NJ, 2002.
9. Hovey, Otis L., *Movable Bridges*, John Wiley & Sons, New York, NY, 1926.
10. Kassimali, Asalm, *Structural Analysis*, 2nd Edition, Brooks/Cole, Pacific Grove, CA, 1999.
11. Mahoney, Timothy R., *River towns in the Great West: The structure of provincial urbanization in the American Midwest*, 1820–1870, Cambridge University Press, Cambridge, UK, 1990.
12. Maltby, F. B., "The Mississippi River Bridges: Historical and Descriptive Sketch of the Bridges over the Mississippi River," *Journal of the Western Society of Engineers*, Vol. VII, No. 4, 1902.

13. McAlpine, C. L., *Report of the Engineer on the Proposed Railroad Bridge Across the Mississippi River at Fulton to Connect the Galena and Chicago Union Railroad with the Chicago, Iowa and Nebraska Railroad, the Lyons, Iowa Central Air-Line Railroad, and the Dubuque, Fulton, and Chicago Railroad,* Daily Press and Tribune Co., Chicago, IL, 1858. Part of this report is included in [19].

14. McGuire, William, Gallagher, Richard, and Ziemian, Ronald, *Matrix Structural Analysis,* John Wiley & Sons, New York, NY, 2000.

15. Michalos, James and Wilson, Edward N., *Structural Mechanics and Analysis,* Macmillan, New York, NY, 1965.

16. Miller, Donald L., *City of the Century: the Epic of Chicago and the Making of America,* Simon & Schuster, New York, NY, 1996.

17. Morse-Kahn, Deborah and Trnka, Joe, *Clinton Iowa: Railroad Town,* Iowa Department of Transportation, Clinton County Historical Society, Clinton, IA, 2003.

18. Nyman, William E., "J.A.L. Waddell's Contributions to Vertical Lift Bridge Design," *Proceedings of the 7th Historic Bridges Conference,* Cleveland, OH, September, 2001, 19–22.

19. Pierson, Joe, *Annual Reports of the Galena & Chicago Union Railroad 1848–1863,* in 2 vols., Chicago & North-Western Historical Society, North Riverside, IL, 2006, unpaged.

20. Ryall, M.J., Parke, G.A.R., and Harding, J.E., *The Manual of Bridge Engineering,* Chapter 12, "Movable Bridges", Thomas Telford, London, UK, 2000, pp. 662–698.

21. Schneider, C. C., "Movable Bridges", *Transactions, ASCE,* Vol. LX, No. 1071, 1908, pp 258–336.

22. Sims, Ronald D., Guhr, Robert C., and Swanson, Paul, "The Mississippi River Bridges at Clinton, Iowa," *North Western Lines,* Elmhurst, IL, Spring, 1993, pp. 24–34.

23. Waddell, John Alexander Low, *Bridge Engineering,* Vols. I and II, John Wiley & Sons, Inc., New York, NY, 1916.

24. Warren, G.K. (Major of Engineers and Bvt. Maj. Gen., USA), "Bridging the Mississippi River between St. Paul, MN, and St. Louis, MO" in Appendix X, *Report of the Chief of Engineers,* Government Printing Office, Washington, DC, 1878.

25. Wilkie, William E., *Dubuque on the Mississippi 1788–1988,* Loras College Press, Center for Dubuque History, Dubuque, IA, 1987, 193.

26. Williams, Clifford D. and Harris, Ernest C., *Structural Design in Metals,* 2nd Edition, The Ronald Press Company, New York, NY, 1957.

3 The Dragon Bridge of Li Chun in Ancient China

Martin P. Burke, Jr. and Huan Cheng Tang

CONTENTS

INTRODUCTION

This open-spandrel, segmental, stone-arch bridge was constructed in Ancient China for and during the Sui Dynasty. It was completed in 606 A.D. after more than a decade of construction and more than two centuries before the Vikings invaded the British Isles. It is known by various names including the Zhaozhou Bridge (after the name of the town when and where it was located), the Great Stone Bridge (in recognition of its long-span achievement), the Dragon Bridge (in recognition of the dragon motif used for its carved railing posts and panels), and the An Ji Bridge (meaning safe crossing). The appearance of this bridge startled and impressed first time observers because of its awesome and seemingly effortless leap across the Xiao River, and the gleaming white limestone surfaces of its façade. Its apparently mystifying presence was also probably enhanced for observers who upon crossing the structure would become surrounded by a panorama of dragon figures in apparent motion on and through the panels and posts of the bridge's parapet type railings. Even a century after its construction, the An Ji Bridge so impressed Zhang Jia Zhen, a local government official of the ruling Tang Dynasty (618 A.D.–906 A.D.), that he felt compelled to describe the bridge for his records. And not only did he praise the bridge's unusual structural characteristics and railing panel carvings, he also immortalized the builder of the bridge by recording his name. Thus the builder of the An Ji Bridge has been identified for posterity as the stonemason, Li Chun. He is one of only a few craftsmen (engineer-architect-artist) of Ancient China who gained dynastic recognition not only for the uniqueness and size of their structures but also for the high quality of their work.

Based on the brief recognition noted above, it would appear that next to nothing is known about the stonemason, Li Chun. But that is not the case. For bridge engineers, architects, artists, sculptors and other craftsmen who have had the opportunity and experience of conceiving, designing and constructing bridges or other structures, an examination of the An Ji Bridge's scientific, technological and aesthetic characteristics, taking into consideration the time, manner and material of its construction, will raise in consciousness a glimmer of Li Chun's genius. That genius, coupled with determination and downright courage, enabled him to make the decisions, the results of which are everywhere evident for recognition, evaluation, and wonder. And that will be the purpose of the following portions of this paper. First to become acquainted with the unusual characteristics of the An Ji Bridge. Second, to consider the thoughts that Li Chun must have had during the conception and construction of this structure. Then to evaluate the decisions he made to achieve what is known by most knowledgeable bridge professionals as one of the most significant early revolutionary structural conceptions in bridge engineering history.

Furthermore, when considering that the remarkable structural innovations incorporated into the An Ji Bridge were also accompanied by outstanding aesthetic characteristics, it becomes quite clear why Li Chun became the subject of myth and legend, and why the An Ji Bridge became the epitome of bridge design excellence. It thus also became an example that was to be emulated in China for many centuries after its completion.

SEGMENTAL ARCH BRIDGES

In his monumental study of *Science and Civilization in China*, Joseph Needham stated,

> [T]he first great segmental arch bridge was constructed by Li [Chun] in China after +600, but no such structures were built elsewhere until Italy followed with several of the kind shortly after +1300, and the design has flourished since. Here the dates correspond well with reports which Italian travelers in Mongo, China could have brought back.[1]

Here, in a carefully qualified statement, Joseph Needham suggested that the segmental design of the Ponte Vecchio (the first segmental arch bridge built in the West) and other similar Italian bridges, and many others that followed elsewhere, were direct reflections of the An Ji Bridge built by Li Chun. From such statements by Needham and others, it would appear that Li Chun originated the use of segmental arch bridges. But careful readers will notice that although Needham acknowledged Li Chun for his unique structural accomplishments, his use of the term "great" appears to be based on his recognition that segmental arch bridges originated in China much earlier than Li Chun's An Ji Bridge.

As apparently recognized by Joseph Needham, any great work of engineering and architecture owes its existence and appearance to the evolution and performance of successful progenitors. And evidence of some of these An Ji Bridge progenitors have been found by Huan Chang Tang, the co-author of this paper. In his book on the *History of Chinese Science and Technology—Bridges* (2), Tang provides photographs of a number of stone panel carvings from the Han Dynasty (202 B.C. to 220 A.D.) depicting the use of short-span segmental arches being used as roadway bridges for small stream and canal crossings (Figure 3.1). However, unlike the An Ji Bridge, which has a shallow rise of only 19 percent of the span length and a relatively straight roadway, these early segmental bridges had little or no approach embankments, arch rises of nearly 25 percent of their span lengths, and roadways that paralleled arch curvatures. Consequently, to negotiate such high rise segmental arch bridges, special practices had to be devised to help horse-drawn carriages cross over the hump of such bridges. And the carved panels from the Han Dynasty illustrate how this was done.

FIGURE 3.1 Images from stone panel carvings from the Han Dynasty (202 B.C. to 220 A.D.) depicting short-span segmental arches being used as roadway bridges for stream and canal crossings.

Some of these carved panels show a group of three laborers stationed at each end of arch type bridges. It appears that the laborers are handling ropes attached to horse-drawn carriages crossing the bridges. Those at the forward end of a bridge help the horses by pulling its carriage to the bridge's apex, while those at the rear are waiting to slow a carriage's descent by pulling on it from behind. Some of the panels depict arch bridges with intermediate supports while other panels show arch bridges without such supports. So it is clear that some of these panels are depicting single-span segmental arch bridges. Since many carriages with multiple horse teams are also depicted on these panels, and groups of laborers with ropes are shown stationed at each end of each of these bridges, it appears clear that in Ancient China methods had been developed to expedite the movement of vehicular traffic across waterways on segmental arch bridges before the beginning of the last millennium.

So Li Chun should be credited not with originating segmental arch bridges. Instead, he should continue to be recognized as the builder of the first long span (great) segmental arch bridge for road traffic. And since his An Ji Bridge has survived most of its predecessors and contemporaries, he is now recognized as the builder of the oldest great segmental arch bridge in the world. But he has an even greater claim to fame since his An Ji Bridge is also recognized as one of the shallowest or boldest segmental stone arches ever constructed. It is also the first major segmental arch bridge that employed the use of open spandrels. Considering that these major structural attributes were incorporated into a bridge with outstanding aesthetic characteristics, it

should be evident that Li Chun has had few bridge design equals in the world's bridge
building profession.

BRIDGE SETTING

To appreciate many of the An Ji Bridge's structural characteristics and its 1400
year survival, something should be known about the site where this record-setting
span was built. The bridge spans the Xiao River about 25 miles (40 km) southeast of Shijiazhang at Zhaoxian. The Xiao River is a stream that originates in the
Taihang Mountains about 60 miles (96 km) to the west of the bridge site. In the
spring, this river is periodically subjected to large flood flows laden with flow-ice
and/or forest debris. The weather of the region is temperate with a yearly rainfall of
about 22 inches (560 mm). Seasonal temperatures range from a low of about –20° F
(–29° C) to a high of about +95° F (+35° C).

China is a country that has regularly experienced earthquakes. For example,
during the last thirty years, there have been more than fifteen significant earthquakes in China, or more than one large quake every other year. There have been
many lesser earthquakes in the region where the An Ji Bridge is located. Figure 3.2

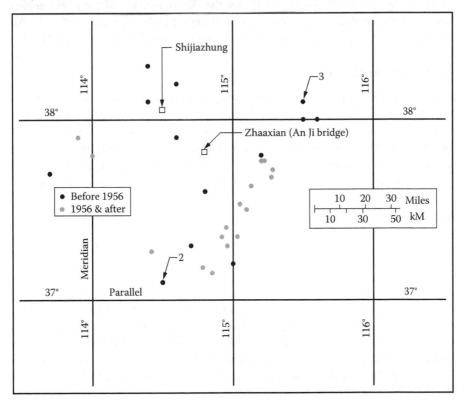

FIGURE 3.2   Known earthquake epicenter locations near the An Ji Bridge. The magnitude
of the earthquakes that occurred between 1956 and 2002 varied from about 4.70 to a maximum of about 7.2 for an arithmetic average of about 5.4.

shows the epicenter locations of these earthquakes. Notice that these epicenters are located within 15 miles (24 km) of the bridge site. Actually, based on the distribution of these epicenters, it appears that there may be an active fault line about 25 miles (40 km) southeast of the bridge. Remarkably, there have been at least sixteen earthquakes in the vicinity of the bridge since it was last rehabilitated in 1956.

BRIDGE DETAILS

BRIDGE TYPE

To provide a suitable structure across the Xiao River, Li Chun avoided the use of river obstructing piers by choosing to build a single long-span stone arch that could span across the entire river. He chose a segmental arch that would be shallow enough to facilitate the movement of road traffic (horses, mules, ox-drawn carts, cattle, foot travelers, etc.) and yet high enough to permit the movement of seasonal river traffic, including oar and pole propelled barges and small sail propelled river craft. That Li Chun's choices for the span (S) of 123'-8⅝" (37.71m) and rise (R) of 23'-8¼" (7.22m) for the An Ji Bridge can be considered bold and courageous is the fact that Spangenberg's Audacity Factor (S²/R) for the An Ji Bridge (640 ft.) was neither equaled nor exceeded in the West for almost eight hundred years.[3, pp. 180 and 181] All structure dimensions given above and below were obtained from Reference No. 4.

SUPERSTRUCTURE WIDTH

It appears that the specter of high flood flows, and the lateral forces associated with such flows, induced Li Chun to provide an extra wide superstructure. At the center of the span the structure measures 30 feet (9.0 m) out-to-out of arch ribs. Obviously, for China in 606 A.D. a structure of that width was substantially more than was necessary to adequately serve road traffic. Additionally, the superstructure was symmetrical and slightly wider at the spring lines than at the center of the span.

ARCH RIBS

One of the most remarkable features of the An Ji Bridge is the huge size of some of the arch-rib voussoirs. The largest of these voussoirs is 3'-4½" (1.03 m) deep, 3'-6⅞" (1.09 m) long, and 1'-3¾" (0.40 m) wide. Consequently, these large voussoirs weigh almost 2,500 pounds or 1¼ tons (11 kN) when finished. In other words, these massive blocks are more than half the height of the average man and more than fifteen times as heavy. What ever possessed Li Chun to make some of these voussoirs so huge and consequently so difficult to handle? Presumably, the depth of voussoirs was based empirically on the successful performance of many earlier segmental arch bridges. But, the length of voussoirs could have been any length to facilitate their handling, transporting, finishing and erecting. Apparently, Li Chun wanted to use as few joints (or as few flaws) in the arch ribs as possible. For this reason alone he may have chosen the longest voussoirs that could be handled by the mechanical devices and animal power available at the time.

Since the limestone for the bridge was taken from a quarry in the hills of Hebei Province, located about 19 miles (30 km) west of the bridge site, Li Chun probably had to build a cart or tow path (including structures for crossing small streams) for that distance suitable for wheeled carts or sleds capable of supporting up to 1½ ton (13 kN) stones, and beasts of burden capable of pulling them. Consequently, reducing the necessary length of voussoirs would considerably reduce their size and weight and thus considerably simplify the quarrying, lifting, transporting, handling, finishing and erection of them. Such reductions in the sizes of these stones would also considerably reduce the number of years that would otherwise be required to build the bridge.

Although the length of arch rib voussoirs may have been chosen to achieve the greatest structural integrity for the ribs, and consequently for the structure itself, aesthetic considerations alone may have dictated the choice of longer voussoirs for the exterior arch ribs. This aspect of the design is discussed more fully hereafter where aesthetic considerations are examined.

ARCHED-STONE-COURSE

Superimposed on the 28 arch ribs is a 1'-1" (0.33 m) thick arched-stone-course (Figure 3.3). This thick course helps to make the superstructure more resistant to lateral flood-flow forces. It also provides longitudinal support for the center walls of the spandrel voids, it serves to improve lateral distribution of superimposed loads, and it probably acts somewhat compositely with the arch ribs to help support super-imposed dead and live loads.

With respect to the arched-stone-course, the present structure contains some peculiar structural details that presumably were not part of the original structure constructed by Li Chun. These details consist of twelve square lateral extensions of the arched-stone-course and five transversely oriented wrought-iron tie rods. These extensions can be clearly seen below the spandrel voids in Figures 3.3 and 3.4. The round heads of tie rods can also be seen along the joint between the arched-stone-course and arch rib. Presumably, early renovation engineers found progressive differential lateral movement of arch ribs with respect to the arched-stone-course due apparently to the freezing of portions of the water saturated spandrel fill. To prevent such movements, small and square, hooked down extensions of the stone course were provided. Presumably, these short easily fractured extensions were later found to be ineffective in resisting the differential lateral movement of the arch ribs, so later renovation engineers installed somewhat more effective wrought-iron tie rods.

SPANDREL VOIDS

It also appears that the specter on high flood flows induced Li Chun to improve the bridge's waterway by providing large voids through the spandrel walls above the arch ribs (Figures 3.3 and 3.4). This innovation in arch bridge design is one of Li Chun's primary claims to fame since the An Ji Bridge is considered to be the world's first open spandrel segmental arch bridge. When examining these voids, it is immediately apparent that Li Chun chose the largest possible voids that would fit between the arched-stone-course at the bottom and the deck-stone-courses at the top. This design

FIGURE 3.3 Close-up view of the arched-stone-course superimposed on the arch ribs of the recently restored An Ji Bridge.

FIGURE 3.4 A recent view of the An Ji Bridge showing the remarkable structural and aesthetic characteristics of the structure.

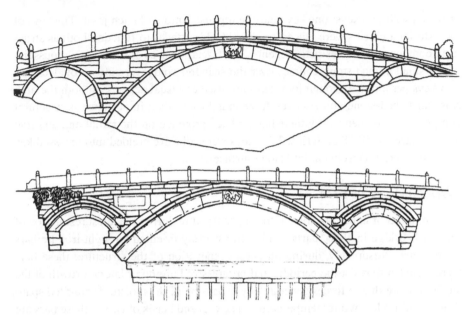

FIGURE 3.5   Fang Shun Bridge, Man Cheng, Hebei Province, China, 309 A.D. (top). Xiao Shang Bridge, Lin Ying, Hebei Province, China, 584 A.D. (bottom).

provided the An Ji Bridge with the greatest waterway, or the least obstruction to major flood flows. That the An Ji Bridge has survived for almost 1400 years attests to the fact that the increased waterway provided by the spandrel voids was an effective design provision for the Xiao River bridge site.

In his study of Ancient Chinese Bridges (2), Tang found and documented the construction dates and design details of many extant segmental arch bridges. These bridges included the 309 A.D. Fang Shun Bridge of Man Cheng (Figure 3.5, top) and the 584 A.D. Xiao Shang Bridge of Lin Ying (Figure 3.5, bottom), both of Hebei Province, China. Both of these bridges *now* have stone-lined voids through their abutment walls. If the bridge details indicated by the sketches of Figure 3.5 do in fact mirror those of the original bridges, then these bridges may have been the inspiration for the An Ji Bridge characteristics. Based upon these considerations, Li Chun probably did not originate the use of stone-lined voids for bridges in China. Instead he merely adopted the use of these voids but moved them to the super-structure where they would not only facilitate flood flows, they would also serve to reduce arch-rib loadings.

CAST-IRON JOINT KEYS

All of the transverse joints of each arch rib are tied together with bow-tie shaped cast-iron ties or keys. Presumably, Li Chun provided these keys to ensure tight-fitting joints and a somewhat more uniform compression of all of the arch ribs in response to the application of superimposed loads. These keys also help to make the super-structure more resistant to the movements and forces associated with flood flows, earthquakes and other environmental changes. One side of each arch-rib joint was provided with two keys; one key near the top and bottom of each joint. The keys of the exterior arch ribs were placed on the outside surfaces to function both as structural components and aesthetic accents. The fascia surfaces of the exterior voussoirs of the small arched-stone-courses over the spandrel voids were also provided with two keys per joint. Although joint keys are used as visual accents on both the arch ribs and arched-stone-courses over the spandrel voids, their importance as structural components is evidenced by their having been provided on the radial and abutting joint surfaces of all 28 arch ribs. Joint keys were adhesive bonded into recessed key ways that were precisely cut into stone surfaces.

WROUGHT-IRON TIE BARS

After removal of upper portions of the superstructure during the latest restoration of the An Ji Bridge in 1956, parts of nine transversely oriented wrought-iron tie bars were found. Presumably, during an earlier renovation of the structure, these bars were installed to resist the periodic and progressive bulging or lateral growth of the superstructure due to forces generated by expansion of the moisture saturated spandrel fill during low winter temperatures. The exposed heads of two of these bars are visible in Figure 3.4. One of these bars is also visible in Figure 3.3. They are located just below and at both ends of the short spandrel walls at the joint between the arched-stone-course and exterior arch ribs. Five other bars are located at the apexes of the arch ribs and the arched slabs over the spandrel voids. Although replicas of these wrought-iron bars have been placed in the restored structure to continue fulfilling their structural purpose of resisting lateral growth of the superstructure, because of their flat, thin and rectangular shape, it is highly unlikely that they represent members of the original An Ji Bridge structure.

FOUNDATIONS

The foundations of the An Ji Bridge are the one aspect of the bridge's construction that has puzzled bridge engineers who have become familiar with its design. For the length of the span and shallowness of the arches, most knowledgeable engineers would expect rather large and deep spread footings to distribute the large thrusts of the shallow arches over the broadest sandy-clay subsoil area possible. Otherwise, progressive longitudinal and vertical movement of the foundations and settlement of the structure would be expected. Nevertheless, the abutment foundations of this bridge are rather small for the size of the structure. Their width, length and thickness are 31'-6" (9.60 m), 19'-0³/₈" (5.80 m), and 5'-0¾" (1.54 m), respectively. In addition, the top of the foundation stones are 6'-6" to 8'-3" (1.98 to 2.51 m) below the ground surface. Surprisingly, these relatively small foundations and their depth from the ground surface to the bottom of footings (11'-9" or 3.58 m) have provided satisfactory service since no noticeable settlement of the structure has been observed. A possible explanation for this satisfactory behavior may be related to the type of falsework that Li Chun used for the erection of the superstructure.

FALSEWORK

The use of 28 individual arch ribs for the An Ji Bridge's superstructure instead of a system of interlocking longitudinal and transverse voussoirs would lead one to suspect that a part-width timber falsework had been used for the construction of the bridge. Such a falsework would minimize the amount of timber that would be needed to erect all of the arch ribs. However, such a falsework would have had several major disadvantages. It would have had to be released and moved laterally for each set of ribs to be erected; it would have made handling, moving and erecting the huge voussoirs difficult; the narrower more flexible falsework structure would have been vulnerable to collapse due to earthquake motions; and since such falsework would almost completely fill the river cross section, it and the partially erected stone work would have been vulnerable to complete collapse during spring flood flows.

On the other hand, the long-term behavior of the structure (little or no settlement or longitudinal translation of the bridge foundations after removal of falsework) would lead one to believe that the actual falsework used by Li Chun consisted of a full-width hand-tamped arched-shaped embankment with a timbered surface. Since such a falsework would have been in place for more than a decade, the foundation subsoils of the bridge also would have been pre-compressed both vertically and horizontally for that period of time. Consequently, such construction could account for the unusual fact that little or no settlement of the structure was observed after falsework removal.

The primary disadvantage of an embankment-type falsework is that the course of the Xiao River would have had to be diverted around the bridge site during construction. But such a procedure would not be as difficult as one might imagine since the shallow bed of the Xiao River is located in an area with rather flat topography. Additionally, it was interesting to learn that the river at the bridge site was de-watered for the 1956 refurbishing of the bridge. Obviously, the main advantage of an embankment type falsework is that the danger due to flood flows and earthquakes would be minimized and movement and assembly of the huge arch-rib voussoirs would be facilitated.

STONE ARCH ANALYSIS

To become better acquainted with the characteristics of the An Ji Bridge and the effects of the spandrel voids on the intensity of the arch thrust, and on the location of the Thrust Line with respect to the intrados and extrados of the arch ribs, a dead-load analysis was made for the arch ribs considering both the presence and absence of spandrel voids. This dead-load analysis was unusual in one respect in that the effect of longitudinal earth pressure was included for the case of the superstructure without spandrel voids. Typically, for the dead-load analysis of segmental arch bridges, longitudinal earth pressures are neglected.

For this analysis, a density of filler material was assumed as 110 pounds per cubic foot (17.28 kN/m³). At-rest earth pressures were assumed to account for initial compaction and to allow for continual compaction due to the movement of roadway traffic. Below is part of the results of that analysis for a 1'-0" (0.30 m) width of arch rib.

Arch Bridge with and without Spandrel Voids, Including Effect of Longitudinal At-Rest Earth Pressure for the Bridge Without Spandrel Voids		
Spandrel Voids	No	Yes
DL Horizontal thrust	128 kips (569 kN)	71 kiks (316 kN)
DL Vertical reaction	114 kips (507 kN)	50 kips (222 kN)

Based on these results, the spandrel voids reduced the DL Horizontal Thrust by 44% and the DL Vertical Reaction by 56%. If longitudinal earth pressure is neglected in the analysis for the bridge without spandrel voids, then the reduction due to the use of spandrel voids equals 28% and 52% respectively. These high reductions are a good indication of how very effective the spandrel voids are in reducing the An Ji Bridge forces and stresses.

The structure was also analyzed for the effect of two 15 ton (133 kN) trucks place side-by-side and located directly over the spandrel voids. For this loading, the Thrust Line within the arch ribs moved towards the introdos of the unloaded half of the span. The structure was safe for this loading since the Thrust Line was located at the edge of the middle half of the rib depth. Also, the unit compressive stress at that location was quite low at about 0.5 ksi (3145 MPa). This stress is considered satisfactory for sound limestone even though some of the limestone voussoirs have been exposed to weathering for 1400 years.

Considering the size of the An Ji Bridge and the science and technology that probably existed at the time of its construction, the conception of this structure and the structural characteristics that were chosen seem almost miraculous. Only a very, very talented bridge builder would have had the intuition and practical experience to construct such an outstanding structure. And when the results of a structural analysis the structure are examined, it becomes clear why the term "genius" has been used to characterize its builder. And that characterization seems appropriate when it is considered that even today, bridge design engineers and architects would be hard pressed to achieve comparable results.

DURABILITY

In a short section on the durability of stone arch bridges, and especially for one like the An Ji Bridge, it is obviously impossible to describe in detail all of the effects that environmental phenomena (movement of roadway traffic, precipitation, daily and seasonal thermal and moisture changes, flood flows, earthquakes, plant root growth, etc.) can and do have on the integrity and durability of such structures. And this is especially true of spandrel filled stone-arch bridges where the fill material that supports the bridge roadway is composed of compacted but permeable soils. But it is known that the An Ji Bridge has been in sorry condition a number of times during its long and eventful life. Some surviving maintenance records indicate that the An Ji Bridge was renovated in 793, 1056, 1063, 1562 and 1597. Presumably, there have been many other renovation projects where work was documented but records lost,

FIGURE 3.6   Elevation view of the An Ji Bridge immediately prior to its 1956 restoration.

or where work was done but not recorded. For example, near the end of the Ming Dynasty (1644 A.D.) the western edge of the superstructure including five arch ribs collapsed. The fallen portion of the structure was restored sometime in the Qing Dynasty (1644–1911 A.D.). Subsequently, the eastern edge of the superstructure collapsed including three arch ribs. The remaining portion of the bridge was modified to survive in that condition until the 1956 restoration (Figure 3.6).

During the study that was made as part of the 1956 restoration, the Xiao River was de-watered so that a search of the river bottom could be made. During the excavations that were made as part of this search, over 1200 bridge stones were recovered. Many of these stones were believed to be from the original structure built by Li Chun at the beginning of the 7th Century. These stones were used as a basis for the restoration. Many of them are now on display in an exhibition house near the bridge site. Some of these displayed stones are believed to be portions of the original fascia arch ribs and dragon-railing panels described by Zhang Jia Zhen, the Tang Dynasty official who wrote about the bridge when it was a little more than a century old.

So although the An Ji Bridge has managed to survive for almost 1400 years, much of the original stone work has been replaced. However, since the latest restoration was based on a study of recovered artifacts believed to be from the original bridge, the structural and aesthetic characteristics shown in Figures 3.3 and 3.4 are believed to be, with a few exceptions, a reasonably accurate replication of the original bridge designed and built by Li Chun. With respect to the structure's primary aesthetic characteristics, some of them are discussed and described below.

AESTHETICS

INTRODUCTION

After choosing the primary structural characteristics of the An Ji Bridge (Roadway Width, Grade, Span Length, Arch Rise, Arch Depth, Spandrel Voids, and Wall Type Abutments), Li Chun provided the structure with continuous deck-stone-courses and continuous railings (First page illustration). Since the wall type abutments and the arch type superstructure have such radically different forms, the continuous deck-stone-courses and the continuous railings were made to extend from the end of one abutment across the superstructure to the end of the other abutment. In this way these three disparate looking elements of the bridge were tied together to give the bridge a semblance of visual unity.

The bridge was provided with relatively plain vertical surfaces that extend from the water surface below the structure to the deck-stone-courses above, and from one end of the bridge to the other. Exposed stone joints were provided with square edges and precisely finished surfaces so that when the joints were tightly fitted together, they would be almost invisible when viewed from a distance (Figure 3.4). White limestone was chosen for the bridge's composition and a relatively flat finish for its surfaces. With respect to the color and texture of the An Ji Bridge surfaces, it appears that Li Chun wanted to achieve an almost surreal or mystical lightness for his extremely long-span arch structure. Consequently, he provided the structure with bright and uniform surfaces relatively free of lines and shadows; surfaces that would glisten in bright sunlight and glow in subdued moonlight. River travelers who were approaching and passing under the An Ji Bridge for the very first time must have been surprised at the vision of this spectacularly light and slender long-span arch making a seeming effortless leap across the width of the entire river from one bank to the other.

Although the An Ji Bridge is composed of thousands of individual stones, they have been so precisely cut and fitted together that the structure appears to be one integrated uniform whole. This aspect is all the more remarkable when it is realized just how much planning, organization, training and control were needed to achieve such a unified result from a crew of variously talented, variously motivated and variously experienced craftsmen working throughout the decade that it took to construct the bridge.

To more fully appreciate the aesthetic accomplishments of Li Chun, it is first necessary to examine some of the major visual aspects of the bridge with a closer focus. In this examination, notice that Li Chun's aesthetics interests and efforts appear to have been directed towards enhancing the visual slenderness and transparency of the bridge's basic structural form, and in drawing attention to the unity and visual quality of its subsidiary structural members.

ARCH RIBS

As noted above under bridge details, the length of exterior arch-rib voussoirs was based on construction, structural and aesthetics considerations. But it appears that the latter consideration had a greater influence on voussoir length than either of the former.

For appearance purposes, the exposed joints between fascia arch-rib voussoirs were provided with square sharp edges and precisely tapered and finished surfaces. Consequently, when these voussoirs were assembled and tightly butted together, their joints would be almost invisible when viewed from a distance (Figures 3.4 and 3.7). So for aesthetic purposes, it would seem that any spacing chosen for these joints would be satisfactory. But since it was intended that these voussoir joints would be provided with cast-iron joint keys, keys that would be highly visible against a white limestone background, even to the casual observer, the position and spacing of these keys would have a significant effect on the appearance of the arch ribs, and on the appearance of the bridge as a whole. Consequently, the appearance of these keys and their location and spacing made them a significant visual feature of the bridge's surfaces. So the problem for Li Chun was how to choose a visually suitable spacing for the joint keys or, in effect, a suitable length for the exterior voussoirs.

Since the voussoirs are 3'-4½" (1.03 m) deep and over 1'-0" (0.30 m) wide, a short length of voussoirs would make them smaller and easier to handle. But a close spacing of joint keys would make them dominate the appearance of the arch ribs. In addition, it would also increase the number of joints that would have to be finished and it would slightly decrease the structural integrity of the arch ribs. On the other hand, a wider spacing of joint keys would improve the appearance of the arch ribs, it would reduce the number of joints that would have to be finished and it would increase the structural integrity of the arch ribs. But, such increased length would make the huge voussoir blocks much more difficult to handle, finish and assemble. With such pros and cons, how was this issue of voussoir lengths decided? A glance at other arch bridges suggest that aesthetics alone guided the choice of voussoir lengths.

Based on the appearance of a number of Ancient Chinese arch bridges, bridges that all bear a striking resemblance to the An Ji Bridge, the longitudinal spacing of voussoir joints was usually made greater than the depth of the arch ribs, but never more than 1½ times the depth. And this observation is also true for the An Ji Bridge. So surprisingly, it appears that the longitudinal spacing chosen of the joint keys determined the size of the voussoirs and the volume or weight of blocks of stone that had to be quarried, handled, finished and erected. Generally, when observers are impressed with the appearance of a structure, rarely are they aware of the thought, effort, time and attention that were devoted to the project details, or the expenses that were incurred just so that the finished structure would have suitable aesthetics characteristics. The construction of the An Ji Bridge is a good example of the results obtained from such considerations. And the exterior voussoirs of this structure are a good example of the cost in time, labor and materials that were involved to produce a work of outstanding structural and visual quality.

To achieve voussoirs that would correspond both geometrically and dimension-ally to the curvature of the arch ribs, the stone masons had to be able to lift and rotate

FIGURE 3.7 Close-up view of the An Ji Bridge's new dragon railing panels and posts. Also visible are exterior arch rib fascia details including; cast-iron joint keys, protruding bead highlight strips, and the carved head of a *Drinking Water Animal*.

these huge rough blocks of stone so that all six sides of the blocks could be precisely cut and finished. For the exterior arch-rib voussoirs, the masons had to provide squared edges and smooth straight or curved surfaces. And remarkably, in reducing the voussoir blocks to provide smooth fascia surfaces, the reduction of the stones had to be accomplished in such a manner that finger-size half-round raised strips parallel to the upper and lower edges of the voussoirs would remain (Figures 3.3 and 3.7). And these finger-size half-round strips had to be so precisely located on the fascia surfaces that when the voussoirs were assembled side-by-side, they would appear like two continuous strips rather than a series of separate pairs of strips. If the pairs of strips on a voussoir were incorrectly located or if they were broken during handling, or if the edges of the finished surface were broken, then the damaged voussoir would have been replaced with another that had been more successfully finished. What is most remarkable about the raised strips on the An Ji Bridge is that instead of a single strip at the top and bottom of the exterior voussoirs, there are actually triple strips at each location (Figure 3.7).

To complete the cutting and finishing of each arch-rib voussoir, precisely located and oriented dove-tail shaped but undercut key ways had to be recessed into the four corners of the accessible vertical voussoir surface. Then, after the voussoirs of each rib were assembled and tightly butted together, the *resulting* bow-tie shaped key ways had to be enlarged to provide a tight fit for the keys. This was done to ensure that arch-rib joints would remain tightly bound together during the application of superimposed dead loads and after exposure to roadway traffic and environmental changes.

It is clearly evident that although the joint keys had the functional purpose of joining the various stones of the longitudinal arch ribs into an integrated structure for the purpose of resisting vertical, longitudinal and lateral forces, their appearance along with the appearance of the raised half-round highlight strips were intended to attract the attention of observers to and along the curvature of the arch. Or in other words, the joint keys and raised strips were intended to reinforce or maximize the appearance of the arch curvature as the primary structural and visual element of the bridge.

In addition to providing carved half-round highlight strips and cut key ways on exterior arch rib voussoirs, a carved animal head was also provided for the central or keystone voussoir. This animal head (Figure 3.7) was intended to represent a spawn of the dragon, a *Drinking Water Animal*. Carvings of these legendary animals were usually placed at the crown of stone-arch bridges. They were believed to possess special supernatural powers to suppress water spirits and control water monsters and thus help to prevent flooding.

Spandrel Voids

Notice that the arched slabs over the spandrel voids were each provided with thin upper "eye-brow" arched-stone-course placed flush with spandrel wall surfaces (Figures 3.3 and 3.4). To produce the thin shadow lines below these courses, the arched slabs and their central supports were slightly recessed behind the fascia surfaces of these courses. And similar to the visual accents provided for the exterior arch ribs, the arched slabs of the spandrel voids were also provided with twin half-round raised highlights strips and two rows of cast-iron joint keys. The upper shadow lines, the two rows of half-round highlights, and the two rows of joint keys were provided by Li Chun not only to enhance and reinforce the curved appearance of the arched slabs, but also to make these slabs match or reflect the appearance of the primary arch ribs. Thus Li Chun used spandrel voids with segmental type arched slabs and with the same visual accents that were used for the primary arch ribs to help visually integrate these disparate structural elements together. He was thus able to achieve a sense of visual unity and uniformity for the An Ji Bridge that would enhance the aesthetic impact of the structure for most observers.

That Li Chun considered the cast-iron joint keys of the *arched slabs* as visual accents rather than structural elements is evidenced by the fact that only the fascia stones of the *arched slabs* were provided with such keys. Notice also that he provided both the large and small arched slabs with the same depth of slab. And that depth was probably chosen not just to satisfy structural purposes, but to provide sufficient depth so that two rows of joint keys (like those used on the main arch ribs) and the two rows of raised highlights would have enough room to be spaced a suitable visual distance apart.

Deck-Stone-Courses

The deck-stone-courses on the sides of the bridge deck were each made to look like one integral and continuous member when viewed from a distance (Figures 3.4 and 3.7). This appearance was accomplished in four ways. First, the various stones of the courses were precisely cut and finished to give them the same thickness and parallel surfaces. Secondly, the courses were made to protrude beyond the fascias of the bridge the same amount at the superstructure and abutments. These protrusions provided uniformly wide and dark shadow bands on the fascias below the courses. Thirdly, the upper edge of the courses were provided with narrow copings to produce narrow shadow bands at the bottom of the copings. Lastly, uniformly spaced circular protrusions were provided on the exposed vertical faces of the courses. Taken together, the courses' uniform thickness, parallel surfaces, upper edge copings, circular protrusions, and uniform shadow bands all coalesced to present to the eyes of distant observers not a series of individual stones but solid, uniform and continuous members providing support for the post and panel type railings.

With respect to bridge railings, both post and panels were continuously embedded in the surface of the deck-stone-courses. This embedment prevented deck drainage from penetrating these junctions and randomly staining the bridge fascias.

But Li Chun was not finished with the aesthetic treatment of the deck-stone-courses. To enhance their appearance at close range, he provided the joints of the courses with square edges and precisely finished surfaces so that the stones could be tightly butted together without shadow lines when assembled. And he had the outside vertical surfaces reduced in the cutting and finishing process to provide not just circular protrusions but decorative rosettes uniformly spaced along the length of the courses. In addition, similar rosettes were also provided on the *upper horizontal exposed outer surfaces* of the courses. Although the rosettes on the vertical surfaces could be seen by all of those viewing the bridge from the river banks or from river craft approaching and moving under the structure, those on the upper surfaces could only be seen by pedestrians walking close to the railings and looking down toward the river below. Finally, to improve the stability of the stones of the courses, they were made from some of the largest stones provided for the bridge. When the design and preparation of the deck-stone-courses are examined in this manner, it should become quite clear that Li Chun must be considered a perfectionist where the appearance of his work is concerned.

However, there is one aspect of the deck-stone-courses that deserves criticism. In Figures 3.3 and 3.4, notice how deep the shadow band is below the deck-stone-course. It is so deep that it partially obscures the narrow shadow bands created by the eye-brow courses over the arched slabs of the spandrel voids. In fact, the shadow band of the deck-stone-course is so deep that it even partially obscures the arch ribs of the spandrel voids. For an aesthetically sensitive designer like Li Chun, this conflict of visual accents appears inconsistent. One explanation for this inconsistency is that the protrusion of the deck-stone-courses beyond the bridge fascias may have been increased during one or more of the bridge's many roadway and railing renovation projects. Since, in these later projects, no attempts were made to replace or replicate the curved dragon railing panels, it would not be surprising to learn that little or no consideration was given to the adverse visual effect that a slight widening of the bridge roadway (and deck-stone-courses) would have on the depth of the fascia shadow bands, and consequently on the overall appearance of the bridge. So, it appears highly likely that the deck-stone-courses provided by Li Chun had shorter protrusions beyond the bridge fascias. Such protrusions would cast much shallower shadow bands that would not conflict with those he provided over the arched slabs of the spandrel voids.

FIGURE 3.8 An original post and railing panel from the An Ji Bridge. Presumably, there would have been four panels like this one with two located in each railing on either side of the bridge center.

Bridge Railing

Prior to the most recent restoration of the An Ji Bridge, all of the railing panels and posts on the structure were of random design and presented a disorderly and unsystematic appearance. Presumably, the original dragon panels and posts described by the Tang government official, Zhang Jia Zhen, were removed or had fallen from the structure and put to other uses. However, when a search of the river bottom was made prior to the 1956 restoration, some railing panels and posts were found that appeared to be those that had been described by the Zhang Jia Zhen. These recovered panels and posts are shown in Figures 3.8 and 3.9. They and other bridge artifacts are now sheltered in an exhibition house near the bridge site. In the 1956 restoration of the structure, the recovered panels and posts served as models for the new railing members. Although the new railing members were made by craftsmen using modern stone carving and finishing equipment, the new dragon panels and posts are considered to be fair simulations of the originals. The original central railing panel carving is shown at the top of Figure 3.9. Its simulated replacement is shown in Figure 3.10. Other new railing panel carvings are also shown in Figure 3.7. In viewing Figures 3.7 and 3.10, notice that dragon railing panel carvings are provided on both sides of the bridge railings. This dual placement is a clear indication that Li Chun was concerned with achieving suitable mythological and aesthetic characteristics for his structure that would be perceived by those viewing the structure from the bridge roadway and by those viewing the structure from the river banks or from river craft.

SUMMARY

The An Ji Bridge is considered to be the first major segmental arch bridge in the world with a span length that was not exceeded in the West for more than seven centuries. It is now the oldest such bridge in the world since the 1400th anniversary

FIGURE 3.9 Original railing panels of the An Ji Bridge photographed almost immediately after they were recovered from the river bottom. Presumably, there would have been two panels like the top panel located at the center of each railing and four of each of the other panels located symmetrical with respect to the center of the bridge.

of its completion was celebrated in 2006. It is also considered to be the first major segmental arch bridge with open spandrels. And this record was not exceeded by a comparable span for over a thousand years. Taken together, these accomplishments place the builder of the An Ji Bridge, Li Chun, in the pantheon of bridge engineers along with other luminaries such as John Roebling, James Eads, Gustav Eiffel, Robert Maillart, Eugene Freyssinet, etc. And when considering the high aesthetic quality of his work, Li Chun also should be considered a structural artist of first rank.

When examining the various unique structural and visual characteristics of the An Ji Bridge, it becomes clear that they were neither accidental or providential. Instead, they appear to be manifestations of human genius; a learned and experienced individual with the structural intuition of the engineer, the spatial recognition of the architect, the visual acuity of the artist, and the form sensitivity of the stone mason. Joseph Needham has made similar observations. In his study, *Science and Civilization in China*, Needham makes a comparison between Chinese, Roman, Medieval and Renaissance Bridges by comparing chronologically their Audacity

FIGURE 3.10   A fair simulation of the An Ji Bridge's original central railing panels.

Factor (Span²/Rise) and their Flatness Ratios (Rise/0.5 Span). Based upon such comparisons, he concluded his study in part as follows:

> It is thus demonstrated that the segmental bridges of +7th-Century China can be placed without hesitation among the best constructions of the kind in +14th-Century Europe; indeed when we allow for [open spandrels, Li Chun] has a priority of more than a millennium, for not until the railway age (the seventies of the nineteenth century) did comparable, if larger, Western bridges arise in the work of engineers such as Paul Sejourne and Robert Maillart. The brilliance of the Chinese anticipation is here made plainly manifest.[3, pp. 180, 182]

Based on the aesthetics speculations given above, it appears that the original An Ji Bridge had a more well defined structural system and a lighter overall appearance than the refurbished structure shown in Figures 3.3, 3.4 and 3.7. Instead of the exterior arch ribs being made flush with the fascias of the arched-stone-course and spandrel walls (as was done in the recent restoration of the bridge, Figure 3.3), the arch ribs probably were recessed a couple of inches (mm) behind the fascias of the arched-stone-course and spandrel walls. This relative position of the exterior-arch ribs with respect the arched-stone-course would have provided the top of the ribs with continuous shadow bands to help reinforce the curved alignment of the ribs. Secondly, the deck-stone-courses probably did not protrude beyond the bridge fascias by more than a few inches (mm) resulting in smaller shadow bands than as shown in Figures 3.3, 3.4 and 3.7. Finally, the small and square extensions of the arched-stone-course and the heads of wrought-iron bars should not be visible on the bridge fascias since it is highly unlikely that these elements were part of the original structure.

When imagining how the An Ji Bridge must have appeared immediately after its construction (First page illustration), it is easy to understand why early bridge observers were so impressed with the appearance of this revolutionary and remarkably slender and transparent bridge. It is no wonder that all of those who were personally able to observe the bridge first hand, sang its praises and began the myths and legends that enveloped the bridge as it continued to serve roadway traffic for the remainder of its millennium and throughout the next.

Although the An Ji Bridge has been known by various names throughout its extremely long life, and although it is presently more generally known as the An Ji Bridge, it could more appropriately be known as the Dragon Bridge of Li Chun. In this respect, the primary focus of Li Chun in the design of the bridge was how best to make the bridge more resistant to the periodic flood flows that the structure would be subjected to in the future. As described above, he provided the longest segmental arch span, open spandrels, a wide superstructure, iron-reinforced arch ribs, and deep foundations. In addition, Li Chun provided the spawn of the dragon at the apex of the arch ribs and he provided brood lairs for dragons in the bridge railing panels (Figures 3.7, 3.8 and 3.9). These latter provisions were entirely consistent with respect to Li Chun's primary purpose, as the Chinese historian, Ann Paludan, has explained,

> The ancient Chinese relied on the mythical power of the dragon to control water spirits and preserve them from disaster.[5]

Aesthetically, the An Ji Bridge was so well conceived that the railing panel carvings chosen by Li Chun were probably not necessary. Nevertheless, considering the time when this structure was constructed, the boldness of the span, and the unique slenderness and transparency of the superstructure, a commonplace less imaginative railing would have been totally out of character with respect to the rest of the structure. Consequently, the dragon motif chosen for the railings not only satisfies cultural expectations but aesthetics considerations as well. Hopefully, this latest restoration of the An Ji Bridge, including the faithful simulation of the original dragon railing panels will reinvigorate the power of the dragons to enable them to protect this structure for many more centuries into the dim and distant future.

Considering all of the attributes of the Dragon Bridge of Li Chun; its great span, its wide roadway, the dimensional precision of its arch voussoirs and other fascia stones, the railing post and panel carvings, and in effect its overall aesthetic characteristics; it appears that this structure was constructed for the use and appreciation by more than just us mere mortals. In this respect, what could have been the motivating force that induced (compelled?) Li Chun to have created such a remarkable structure? The more that this bridge is studied, the more questions it provokes. Unfortunately, after 1400 years, all one can do is to question, to marvel, and to wonder.

ACKNOWLEDGMENTS

The authors wish to express their appreciation to Professor Ronald G. Knapp of SUNY, New Paltz for Figures 3.4 and 3.10 and the original photograph for Page 1 that was digitally altered by the authors especially for this paper; Professor Lichu

Fan, Tongji University, Shanghai, for the seismic data used to prepare Figure 3.2; and Ling Yang for Figures 3.3 and 3.7.

REFERENCES

1. Needham, J., *Science and Civilization in China*, Cambridge University Press London, United Kingdom, 1965, Vol. 1, p. 230.
2. Tang, H. C., *History of Chinese Science and Technology—Bridges*, China Science Publishing House, Beijing, China, 2000, p. 275. (In Chinese)
3. Needham, J., W. Ling and L. Gwei-Djen, *Science and Civilization in China*, Cambridge University Press, London, United Kingdom, 1971, Vol. 4, Part III.
4. Knapp, R.G., Bridge on the River Xiao, *Archaeology* Archaeological Institute of America, Boston, Massachusetts, 1988, Vol. 41, No. 1, p. 54.
5. Paludan, A., *The Reign-By-Reign Record of the Rulers of Imperial China*, Thames and Hudson, New York, New York, 1998, p. 84.

Part 2

Management

4 Bridging the Gap
Connecting Design and Historic Preservation Goals on Milwaukee County's Historic Parkway Bridges

Amy R. Squitieri and Bob S. Newbery

CONTENTS

EXECUTIVE SUMMARY

A historic parkway composed of integrated natural and designed features presents a special challenge to preservationists and engineers. Milwaukee's "Emerald Necklace" is comprised of thirteen early twentieth-century parkways that form an interlaced chain around the city. Contributing to the aesthetic of this necklace of parks and parkways are dozens of bridges built in the 1920s and 1930s using local materials and handcrafted finishes on railings, abutments, and spandrel walls. Early on, historians and engineers agreed on a goal to preserve this rustic design aesthetic of the parkway landscape, but struggled through several bridge replacement projects before ultimately finding an approach to new bridge design that appropriately balanced modern design standards with aesthetic interpretation. This approach is informed by existing guidelines for historic bridge rehabilitation projects but expands beyond these to consider the parkways in their entirety, not just the individual historic bridge. In late 2006, Milwaukee County, the owner of most of the bridges, agreed to follow a new approach to designing replacement bridges that better interprets the historic context of the parkways.

INTRODUCTION

Since the 1960s, the confluence of two forces has led to a revolution in the treatment of old bridges. The 1967 collapse of the Silver Bridge over the Ohio River, between West Virginia and Ohio, made engineers, politicians and many citizens aware of serious deficiencies in bridges on public roads across the country. This recognition led to the adoption of a systematic bridge inspection program to monitor the condition of existing bridges. As summarized by Abba Lichtenstein in his article explaining the collapse of the Silver Bridge and subsequent federal developments for the *Journal of Performance of Constructed Facilities*, "In addition to providing the country with inspection standards, Congress also began to appropriate funding for bridge inspections, rehabilitation, and replacement."[1] For many in state transportation agencies and among the broader public, replacing old bridges could not happen fast enough.

Nearly concurrently, historic preservation was becoming a stronger social movement, pushed forward with the adoption of the National Historic Preservation Act of 1966, which affirmed a federal responsibility for protecting historic properties. The act authorized the Secretary of the Interior—in practice, the National Park Service—to "expand and maintain a national register of districts, sites, buildings, structures, and objects significant in American history, architecture, archeology, and culture" and to provide limited grants for historic properties. As described in the National Park Service's own history, "In specific response to the destruction wrought by federal projects, Section 106 of the act ordered federal agencies to consider the effects of their undertakings on National Register properties and permit the Advisory Council on Historic Preservation to comment on such undertakings."[2]

Meanwhile, a smaller but passionate group of historic preservation advocates were becoming historic bridge enthusiasts. They were discovering that these bridges were not just old and worn, but were historic and had much to tell about developments and trends in history and engineering. As noted in the 2003 retrospective, *Historic Bridges: A Heritage at Risk*, "Historic bridges also are the single most visible icon of highways and civil engineering art."[3] The professional and advocative interests in preserving historic bridges and the pressure to replace them are often resolved through the Section 106 process.

Driven by federal requirements and prodded by bridge enthusiasts, bridge engineers began pushing the limits of rehabilitation and restoration for deficient bridges. From non-destructive testing to creative, cost-saving construction solutions, to capacity enhancements that do not damage the historic fabric, the challenges of keeping historic bridges in use have been explored, documented, and shared.[4] For their part, historic bridge enthusiasts scoured the engineering literature, textbooks, and patent records. They directed historic bridge inventories, a nationwide effort that is now "nearing completion," according to surveys of state transportation agencies.[3] Relocation of historic bridges to less demanding sites became more frequent and the means for transport bordered on the ingenious. The Calhoun County Historic Bridge Park in Calhoun, Michigan, represents a remarkable commitment to relocating and restoring historic bridges and the construction techniques associated with them. Success stories for relocating bridges have ranged from welding wheels onto the

FIGURE 4.1   The Manchester Street Bridge in Baraboo, Wisconsin. This bridge was successfully relocated and is featured on the FHWA's Historic Bridge web page at http://www.environment.fhwa.dot.gov/histpres/bridges.asp. (Photo by author.)

floor beams (turning the truss bridge into a trailer), to using an Army National Guard engineering unit and its truck dollies and tank-puller, to an airlift by Air National Guard helicopter.[5,6,7]* Covered wooden bridges have long had their advocates and protectors, but they too benefited in very practical ways from the surge of interest in historic bridges programs, as evidenced by the National Historic Covered Bridge Preservation (NHCBP) Program.†

A number of specific issues surrounding continued use of historic bridges have been well covered when the bridge is the historic resource and focus of the preservation effort. The basic tension that arises between engineering and preservation values, with cost effectiveness lurking in the background, is now familiar. This paper focuses on a different issue: where the bridge is a part of a larger historic district and the overall aesthetic and historic value of that district is the focus for the historic preservation community. In such circumstances, the fight to save the bridge may be more readily conceded.

For Milwaukee County's historic parkways, the challenge lies not in whether or not to preserve individual bridges, but in how to maintain the historic integrity and significance of the entire historic district. To address this, engineers and historians turn to the guidance offered by the Secretary of the Interior's *Standards for Treatment of Historic Properties with Guidelines for Preserving, Rehabilitating, Restoring and Reconstructing Historic Buildings* (Secretary of the Interior's Standards).[8] Recent projects on Milwaukee County's historic parkways serve to illustrate the potential difficulties in interpreting the aesthetic guidance provided by the Secretary of the

---

* Bridge No. 4846 in Minnesota was the first known to the authors to use the welding of wheels to the truss, see SIA *Newsletter*, Vol. 13, No. 3 & 4, Fall & Winter, 1984, p. 6. Correspondence from Mn/DOT, March 20, 1986, in personal files of Robert Newbery. The Army National Guard was called upon by Dodge County, Wisconsin, to assist in moving the Ninabuck Road Bridge. According to the caption for a photograph in *The New York Times,* March 9, 2006, *"The Iowa Air National Guard moved the three spans of the historic Hale Bridge yesterday by Chinook helicopters to a new site..."*
† This program was established by Section 1224 of the Transportation Equity Act for the 21st Century (TEA21).

Interior's Standards. One facet of this case is the resolution of a common safety problem faced with historic bridges: providing crash-worthy railings.[9] This issue is heightened on the parkway bridges as much of the design aesthetic is articulated in the railing designs, which use local materials and handcrafted finishes in keeping with the natural environment.

HISTORIC PARKWAYS IN MILWAUKEE COUNTY

Preservationists and urban planners are increasingly recognizing parkways, many built in the 1920s and 1930s, as significant historic properties that are eligible for listing in the National Register of Historic Places (National Register). With this recognition, a parkway receives consideration under Section 106 when a federal undertaking that affects all or part of a parkway is proposed by a local public works department or state transportation agency. As the Section 106 process unfolds, historians and engineers strive to balance the interests of preservation and engineering. The preservation perspective is informed by the particular significance of parkways as designed landscapes, where each bridge is an integral part of the overall historic resource. The example of Milwaukee County's thirteen, interlaced parkways forming an "Emerald Necklace" is illustrative of national trends in parkway development that unfolded during the late nineteenth and early twentieth centuries.

Parkways and other unified networks of landscaped thoroughfares were introduced in city planning schemes in cities such as New York, Chicago, Minneapolis, and Boston late in the nineteenth century. These efforts, led by landscape architects such as Frederick Law Olmsted, soon converged with those of Progressive Movement advocates who sought to improve social conditions in urban neighborhoods. In the first decades of the twentieth century, parks and parkways became a means of providing a moral and healthy outlet from the city's social ills tied to poverty and crowding.

Milwaukee first joined this national movement when, under direction of a new parks commission, the city and county began to purchase land for parks in the early 1900s.[10] The commission's greatest role in early parkway development came about with its ambitious parkway study in 1923, led by landscaper and commissioner, Charles B. Whitnall. Often referred to as the "Father of the Milwaukee County Park System," Whitnall's philosophies on park influences, design, and social reform guided Milwaukee County's parkway design. He was a major advocate of widespread green spaces because of their beneficial influences on the health and social conditions of the city. He believed urbanization, and the resulting congestion and pollution of the city, separated man from nature.

Decrying the destruction and waste of natural resources as a "public tragedy if not a public crime," Whitnall offered a vision of Milwaukee penetrated and encircled by landscaped rivers and creeks.[11] His initial vision for a county parkway system included 84 miles of parkways that followed the Milwaukee, Menomonee, Kinnickinnic, and Root rivers and their associated creeks, twice encircling the city and county and resulting in a "necklace of green."[12] Each parkway would conform to the natural topography of its respective watershed feature, providing numerous locations for aesthetic features such as retaining walls and bridges to contribute to the

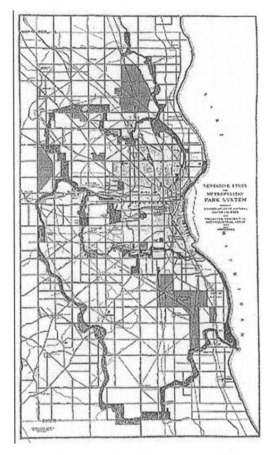

FIGURE 4.2 Tentative county parkway system. (From Milwaukee County Regional Planning Department, *Second Annual Report* (Milwaukee, Wisc.: Milwaukee County Regional Planning Department, 1925, 10.)

parkway experience. The bridges interplay with the natural environment in their use of local materials and rustic aesthetic conveyed in timber and rock-faced, random ashlar stone veneers. Although the structural types vary (concrete arch, slab, concrete girder, and rigid frame types are all found), application of a common stone veneer to abutments or spandrel walls provides a unified aesthetic.

In Milwaukee County, most of the parkway work was completed during the Depression era. As a result of New Deal programs introduced in 1933 by President Franklin D. Roosevelt to promote economic recovery, significant amounts of funding were available for labor and materials for urban projects.

Depression era work relief projects, particularly of the Civil Conservation Corps (CCC) and Works Progress Administration (WPA), proved pivotal to the development of the Milwaukee County park system. This system was in a good position to receive federal money for work-relief projects as a result of the planning and preparations that occurred in prior decades.[13] The majority of federal funding went to CCC

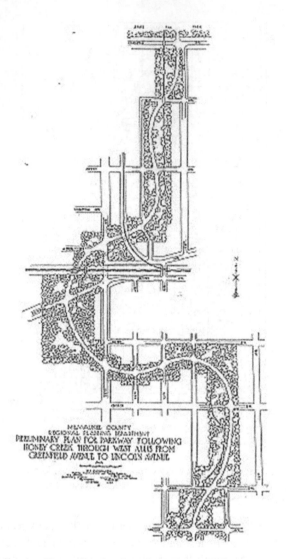

FIGURE 4.3   Following Honey Creek—from State Fair Park through West Allis to Lincoln Avenue. (From *The First Plans for a Parkway System for Milwaukee County: Illustrations from First Annual Report*. Milwaukee, Wisc.: Milwaukee County Regional Planning Department, 1924, 1.)

and WPA laborers to construct roads, bridges, and buildings; develop erosion and flood control measures; and implement landscaping for the park system.[14] The WPA operated a limestone quarry and used the stone to build benches, retaining walls, and bridges throughout the parkways.[15]

Milwaukee County parkways are historically significant under National Register *Criterion A* for their association with events that have made a significant contribution in the areas of community planning and development and federal aid work relief. Additionally, the parkways are significant under *Criterion C* as good examples of

FIGURE 4.4 Largely built as part of a Depression-era construction program, Milwaukee County's parkways contain more than 50 bridges and dozens of buildings, such as this warming house, that reflect the efforts of federal relief laborers through their use of local material, rustic style, and handcraftsmanship. (Photo by author.)

FIGURE 4.5 The West Forest Home Avenue Bridge (B40-979) is a 1931 concrete T-beam bridge with a stone-faced headwall that features a segmental arch opening. Its handcrafted, rock-faced, limestone parapet walls feature a random ashlar pattern. (Photo by author.)

designed landscapes that contain topographic and natural features as well as buildings, structures, and recreational facilities. Further, components of the parkways feature highly-skilled craftsmanship and use local materials to achieve a unified rustic design aesthetic. Components that contribute to the overall design landscape include culverts, bridges and retaining walls with rock-faced, ashlar stone veneers and often timber railings, and parkway buildings such as comfort stations, picnic shelters, and maintenance buildings that use similar natural materials. Together these features convey a unified rustic design aesthetic that gives the parkways their historic character.

DESIGN APPROACH FOR COUNTY PARKWAY BRIDGES

This sense of Milwaukee County's historic parkways as designed landscapes composed of integrated features provides a framework for describing the specific effort to balance preservation and engineering interests that occurred when bridge replacement projects were proposed. This paper presents an approach to preservation adopted in Milwaukee County whereby deteriorated bridges are replaced but the overall design aesthetic of the parkways is retained. This preservation approach focuses on the overall resource—those thirteen early twentieth-century parkways that form an interlaced chain around the City of Milwaukee, known as the "Emerald Necklace."

In this example, replacing a deteriorated bridge is generally conceded and the focus shifts to saving the historic parkways. Milwaukee County, which owns many of the bridges, acknowledged the need for a new approach after the first of its thirteen Depression-era parkways was determined eligible for listing in the National Register. At first, engineers struggled to design bridges that were compatible with the overall aesthetic of the historic parkway landscape. Historians had difficulty understanding how modern engineering standards influence bridge design and articulating useful direction to guide the design process. New direction was needed for current and future projects.

This story starts with the replacement in 2000 of one bridge in Milwaukee County's parkway system. The Oak Creek Bridge (P40-559), originally built in 1931, was replaced in 2000 without rigorous consideration of its historic value within the Section 106 review process.* Moreover, and more importantly, the historic significance and associated aesthetic value of the surrounding parkway as a designed landscape was also overlooked. Because the bridge was listed in the Department of Transportation's bridge database as a non-historic type, and the parkway was not at the time identified as a historic property eligible for the National Register, this replacement project did not trip any alarms and proceeded with perfunctory, programmatic review. The project sponsors were, however, aware in a general way of the aesthetic value of the parkway's landscape and the bridge's role within. Thus, they made an attempt at an aesthetic treatment within the parameters of accepted engineering standards. The designers of the new bridge engaged in lengthy discussions with the state's bridge engineers about an appropriate railing for a low-volume, low-speed bridge in a parkway setting. Current standards called for a solid railing, and the resulting design, featuring steel railings, was less than sympathetic to the original aesthetics of the Oak Creek Parkway.

Within a year or two of this project, additional bridge replacement projects were proposed for several other parkways in Milwaukee County's interlaced system. Though a preliminary study by the Milwaukee Metropolitan Sewerage District had recommended 12 individual parkway bridges eligible for the National Register, these evaluations were never formalized and the parkways themselves had not been evaluated.

---

* Section 106 requires federal agencies to "take into account" the effect of their projects on historic properties. A part of the Section 106 Review Process is to make a good faith effort to identify and evaluate potential historic properties. The details of the process in Wisconsin as it relates to the bridge replacement projects on Milwaukee parkways are beyond the purview of this paper.

FIGURE 4.6 In an effort to match the original design and materials, the new Oak Creek Bridge (P40-559) featured steel railings embedded in veneered posts and stone-veneered abutments. However, the modern styling of the multiple railings and posts contrasts with the simplicity of its historic, rustic design. (Photo by author.)

When the District pursued a major flood control project adjacent to a county parkway, planners, engineers, and historians quickly recognized a common theme. Each of these projects was a federal undertaking under Section 106. As Milwaukee County began to navigate its regulatory requirements, they soon recognized the interconnectedness of these projects and the need for a comprehensive approach.

Engineers unfamiliar with the Secretary of the Interior's Standards for historic properties, and unacquainted with their applicability to the bridges within historic parkways, were faced with defending the recently built Oak Creek Bridge as looking "just like" its predecessor. As articulated by the county's Department of Parks, Recreation and Culture in a letter to the state Department of Transportation, the new bridge on Oak Creek Parkway "replicated the original bridge stone character of the bridges of the parkway...By matching the existing stone material and the masonry design of the existing structures the character of the original landscape design will be preserved. The designers and engineers for this project are implementing aesthetic details and treatments that will help unify the landscape detailing of the parkway."[16] At first, they insisted that other bridges in the parkways be allowed to proceed after the same fashion as the Oak Creek Parkway Bridge. The designers and the local public works department were no doubt unhappy that state agencies, in processing Section 106 paperwork, had now discovered the potential significance of the parkway.

When five new parkway bridge projects were added to the county's upcoming construction program, historians and engineers worked to advise the county of its options. Historians went to work on researching and documenting the county's parkway system to establish its eligibility for listing in the National Register. Engineers aimed for compatible designs that respected the rustic design aesthetic of the parkways. However, modern railings were markedly different than the originals,

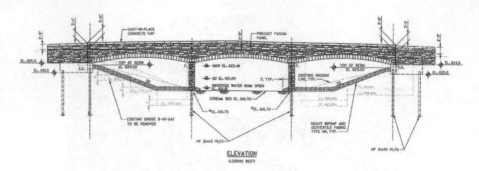

ELEVATION
(LOOKING WEST)

FIGURE 4.7 The design for a replacement of the Milwaukee River Parkway Bridge uses custom form liner to match the original stone veneer. Specifications require mock-up panels of the architectural surface treatment to be approved by the county and application of multicolor concrete stain. (Plan by Collins Engineers.)

form liner was suggested in place of local limestone, and the final product was not aesthetically satisfying. Some preservationists mulled the need for safety improvements, arguing that the existing 2-inch by 12-inch timber railing was appropriate. In the meantime, the replacement of another parkway bridge was allowed to proceed, but it was clear that a gap remained between aesthetic considerations for the historic parkways and modern engineering standards. All parties struggled to defend their different positions before they ultimately came together to embrace a design approach that considered the need for modern design while respecting the spirit of the historic "Emerald Necklace."

DESIGN GUIDELINES FOR PROJECTS AFFECTING HISTORIC PROPERTIES

The design approach that was adopted for new construction on Milwaukee County's historic parkways followed the framework offered by the Secretary of the Interior's Standards. The Secretary of the Interior's *Standards for Treatment of Historic Properties* offer guidance for rehabilitation of historic properties and must be followed when undertaking historic preservation projects using federal funds. A critical definition that informs the design approach is this: "Rehabilitation is defined as the act or process of making possible a compatible use for a property through repair, alterations, and additions while preserving those portions or features which convey its historical, cultural, or architectural values." For the parkways, the "compatible use" is an expansion of their original 1920s function, in that instead of serving solely leisure drivers, the parkways now also carry commuters and bicyclists.

Recognizing that the Secretary of the Interior's Standards were developed for historic buildings and that they had severe limitations when applied to bridge rehabilitation projects, the Virginia Transportation Research Council, in 2001, introduced its own version of appropriate treatments for use with historic bridges. The Virginia Standards have nine recommendations that are specifically designed to consider the unique characteristics of bridges and bridge rehabilitation projects. Among the guidance provided is the following:

All bridges shall be recognized as products of their own time. Alterations that have no historical basis and that seek to create a false historical appearance shall not be undertaken.

Distinctive engineering and stylistic features, finishes, and construction techniques or examples of craftsmanship that characterize an historic property shall be preserved.

Where the severity of deterioration requires replacement of a distinctive element, the new element should match the old in design, texture, and other visual qualities and where possible, materials. Replacement of missing features shall be substantiated by documentary, physical, or pictorial evidence.

New additions, exterior alterations, structural reinforcements, or related new construction shall not destroy historic materials that characterize the property. The new work shall be differentiated from the old and shall be compatible with the massing, size, scale, and architectural features to protect the historic integrity of the property and its environment.[17]

These standards are useful for most historic bridge rehabilitation projects but fall short for Milwaukee County's historic parkways bridges where the entire parkway system is the significant resource and each bridge is an integral part of this designed landscape. To address the broader context for the parkways, the National Park Service's *Guidelines for Cultural Landscape Districts* offer additional direction. These guidelines address issues of importance in a historic landscape including spatial organization, circulation patterns, vegetation, and associated structures. The guidelines for features associated with circulation can inform an approach to rehabilitation of a historic parkway. Specifically, when alterations or additions are required, the guidelines recommend "designing and installing compatible new circulation features when required by the new use to assure the preservation of historic character of the landscape."[18] Furthermore, for new structures in cultural landscapes, the recommendations call for "a new design that is compatible with the historic character of the landscape."[19] This guidance informs the solution for Milwaukee County's parkway bridges, where a more functional structure that accommodates current usage is sought to replace an existing deteriorated bridge.

MILWAUKEE COUNTY'S NEW DESIGN APPROACH

Once Milwaukee County, which owns many of the bridges, acknowledged the need for a new approach, the focus shifted to preserving the characteristics of the historic parkways. Under the county's new approach, engineers and historians are working together to preserve the historic significance of these designed landscapes by maintaining the rustic aesthetic that was part of each bridge's original design. The Section 106 process, coupled with associated agency reviews, provided the mechanism for working out the details of the design approach.

In the Memorandum of Agreement, executed in early 2006, Milwaukee County agreed to incorporate State Historic Preservation Office (SHPO) guidance on bridge designs for five projects.[20] The projects were required to comply with the Secretary of the Interior's Standards and with current bridge design standards. While the Virginia Standards and *Guidelines for Cultural Landscape Districts* were informative, further development was required to find the particular solution for the county's historic parkway bridges. This guidance was determined among SHPO, the Wisconsin Department of Transportation (WisDOT), Milwaukee County, and its engineering and historical consultants during a site visit to the five bridges. Specific design guidance for the new bridges, as determined in the field, was formally incorporated into the Memorandum of Agreement. For each bridge, it is stipulated to maintain the rustic aesthetic of the original bridge while accommodating current standards. The stipulations read as follows:

1. Oak Creek Parkway Bridges—In keeping with the rustic setting and bridge features, the end wall of the new bridge should be longer, extending for the length of the abutment. The end wall should frame a simple railing of either timber or painted steel. Due to design standards, the railing needs to be carried onto the nearest part of the end walls. The railing posts should not be encased in stone. Stone-facing should be added to the sides of the abutments and pier. Existing stone should be reused if possible. Otherwise, it should match the existing stone.
2. Honey Creek Parkway Bridges—In keeping with the parkway's elegant design that fits with surrounding residential areas, the new bridges will have a solid parapet. The height should be raised on the ends (about 6 inches) to frame the bridge. The side elevation should have a finished appearance. Existing stone should be reused if possible; otherwise, it should match the existing.
3. Milwaukee River Parkway Bridges—In keeping with the massive proportions and sweeping design of these bridges and the parkways' wide vistas, the new bridges will have a solid parapet. The height of the rail should peak at the center. The form-liner used on the bridge's exterior surfaces should match the proportions and finish of the existing stone. The side elevation should have a finished appearance, including facing on the piers and abutments. The proposed steel railing and horizontal and vertical frames in the form-liner should not be used. If a railing is required on top of the parapet, it should use a simple design.[20]

A specific review was required to confirm compliance with the articulated design requirements. This process enabled Milwaukee County to proceed with projects as follows:

1. Prior to construction letting, Milwaukee County will provide plans to SHPO to demonstrate that the designs for projects are consistent with SHPO guidance.
2. If SHPO finds that the bridge designs do not comply with the guidance, SHPO will respond in writing with suggested modifications within 20 working days.
3. Milwaukee County will implement suggested modifications or consult to resolve the dispute in accordance with the provision below.
4. If no response is received from SHPO, Milwaukee County will implement the designs as submitted.

Two bridges completed on the Honey Creek Parkway in 2007 illustrate how design details were executed in compliance with design requirements articulated in the Memorandum of Agreement. The construction specifications called for a 4-inch

FIGURE 4.8   The Milwaukee River Bridge (B40-647) displays deteriorated abutment walls and spalling stonework. With a sufficiency rating of 34.7, it is scheduled for replacement in 2008. (Photo by author.)

FIGURE 4.9   The new parapet wall for this Honey Creek Parkway Bridge was substantially taller than the original to comply with current standards but replicates the historic stone veneer. Photographs of the original stonework were used as the reference for matching the new stone masonry. (Photo by author.)

stone veneer with the stone appearance, color, and placement specified to match the original stone parapet in size, shape, color, shades, joint finish, mortar color, and construction. Tradesmen experienced in stone construction were required to perform this work. A 6-inch cap stone was required on top of the parapets to match the original cap stones in length, size, and color.

CONCLUSION

Milwaukee County's historic parkway system is largely intact today; however, the roadways, parks, and related components are constantly evolving to meet the needs of modern society. The development of recent bridge projects on these parkways illustrates the evolution that has transpired as historians, planners, and engineers found a new approach that considers the need for both modern design and historic preservation. The new procedure requires review of bridge designs for compatibility with the historic environs of the parkway and allows compromise with materials, geometrics, and railing design, where necessitated by safety and cost efficiency. This approach is informed by the Secretary of the Interior's Standards and associated guidelines, but expands the considerations to the parkways in their entirety. In order to retain their historic significance and convey the spirit of the "Emerald Necklace," the parkways should retain enough of their characteristic features after bridges are replaced to be clearly recognizable as historic designed landscapes. The adopted approach gives Milwaukee County a clear process to replace deteriorated parkway bridges while retaining the qualities of the parkways that enable them to convey historic significance. In finding this solution, bridge owners, engineers, historians, and agencies learned to trust in each other's expertise, recognize the inherent conflicts, and work together to create a new bridge that meets current engineering standards and yet expresses the original aesthetic intent.

REFERENCES

1. Lichtenstein, Abba, "The Silver Bridge Collapse," *Journal of Performance of Constructed Facilities* 7, no. 4 (Nov. 1993): 249–61.
2. Mackintosh, Barry, *The National Historic Preservation Act and The National Park Service: A History* (Washington, D.C.: History Division, National Park Service, Department of the Interior, 1986). Also available online at http://www.nps.gov/history/history/online_books/mackintosh5/preface.htm.
3. DeLony, Eric. and Klein, Terry H., *Historic Bridges: A Heritage At Risk. A Report on a Workshop on the Preservation and Management of Historic Bridges, Washington, D.C. December 3–4, 2003.* (Rio Rancho, New Mexico: SRI Foundation, 2004), 1. Also available online at http://www.srifoundation.org/research.html.
4. Most recently addressed in *NCHRP Project 25-25 Task 19: Guidelines for Historic Bridge Rehabilitation and Replacement*, March 2007, prepared by Lichtenstein Consulting Engineers, Inc. (reprinted at http://www.trb.org/NotesDocs/25-25(19)_FR.pdf).
5. Caption for photograph in *The New York Times,* March 9, 2006, *"The Iowa Air National Guard moved the three spans of the historic Hale Bridge yesterday by Chinook helicopters to a new site..."*
6. Correspondence from Mn/DOT, March 20, 1986, in personal files of Robert Newbery. The Army National Guard was called upon by Dodge County, Wisconsin, to assist in moving the Ninabuck Road Bridge.
7. SIA *Newsletter*, Vol. 13, No. 3 & 4, Fall & Winter, 1984, p. 6.
8. U.S. Department of the Interior, *Standards for Treatment of Historic Properties with Guidelines for Preserving, Rehabilitating, Restoring and Reconstructing Historic Buildings* (Washington D.C.: National Park Service, 1995). Also available online at http://www.nps.gov/history/hps/tps/standguide/.
9. Harshbarger, J. Patrick, et al., *Guidelines for Historic Bridge Rehabilitation and Replacement*, A-25, Transportation Research Board, March, 2007.
10. Whitnall, Charles B., "How a Lecture Course Saved the Shores of Milwaukee County," The American City [Reprint], n.p.; Charles B. Whitnall, "The Philosophy, Evolution, and Objective of the Milwaukee County Park Commission," n.p.

11. Whitnall, Charles B., "Report on the Milwaukee Metropolitan Park Commission," [Milwaukee], n.d., 8.

12. Milwaukee County Regional Planning Department, First Annual Report. (Milwaukee: Milwaukee County Regional Planning Department, 1924), 22–23.

13. Anderson, Harry H., "Recreation Entertainment, and Open Space: Park Traditions in Milwaukee County," in Trading Post to Metropolis: Milwaukee County's First 150 Years, ed. Ralph A. Aderman (Milwaukee, Wisc.: Milwaukee County Historical Society, 1987), 281–292; Eugene A. Howard, "Recollections on Development of Milwaukee County Park System," N.p., 15.

14. Milwaukee County Regional Planning Department and Milwaukee County Park Commission, Quadrennial Report: 1933–1936. (Milwaukee, Wisc.: Court House, 1937), 49; Milwaukee County Park Commission and Milwaukee County Regional Planning Board, Quadredecennial Report: 1937–1950 Inclusive, 65–72.

15. Milwaukee County Park Commission and Milwaukee County Regional Planning Board, Quadredecennial Report: 1937–1950 Inclusive.

16. Department of Parks, Recreation and Culture, Milwaukee County, correspondence to WisDOT, November 6, 2000, in personal files of Robert Newbery.

17. Miller, Ann, et al., *A Management Plan for Historic Bridges in Virginia* (Charlottesville, Va.: Virginia Transportation Research Council, 2001). Standards excerpted from the complete list of nine.

18. U.S. Department of the Interior, *Guidelines for Cultural Landscape Districts* (Washington D.C.: National Park Service, 1996), 72.

19. U.S. Department of the Interior, *Guidelines for Cultural Landscape Districts* (Washington D.C.: National Park Service, 1996), 81.

20. Memorandum of Agreement Between the Federal Highway Administration and the Wisconsin State Historic Preservation Officer prepared pursuant to 36 CFR § 800.6(c) regarding WisDOT ID # 2987-08-00, Oak Creek Parkway Bridge (P-40-0741) Over Oak Creek, Milwaukee County and WisDOT ID # 2984-12-00, Honey Creek Parkway Bridge (P-40-0779) Over Honey Creek, Milwaukee County and WisDOT ID # 2984-12-02, Honey Creek Parkway Bridge (P-40-0780) Over Honey Creek, Milwaukee County and WisDOT ID # 2099-00-00, Milwaukee River Parkway Bridge (B-40-0647) over Milwaukee River, Milwaukee County and WisDOT ID # 2967-04-01, Milwaukee River Parkway Bridge (B-40-0646) Over Milwaukee River, Milwaukee County (executed on January 19, 2006).

5 Managing Historic Bridges in Minnesota

The Historian and the Engineer Collaborate

Robert M. Frame III and Steven A. Olson

CONTENTS

INTRODUCTION

Since the early decades of the twentieth century, bridge engineers have learned to get along with government oversight and professional codes and standards. Beginning with the federal Bureau of Public Roads, the first state-aid act, continuing through the successive formation of each state highway commission, and extending into the American Association of State Highway and Transportation Officials, engineers have accommodated themselves to institutionalized standards, guidelines, and reviews of their work.

The arrangement was largely routine and predictable until 1966 when President Lyndon Johnson signed into law the National Historic Preservation Act (NHPA).[1] This act vastly widened the involvement of the federal government, and subsequently the state governments, in the protection of the nation's historic resources. Simultaneously, the Act inserted cultural resource management, in the form of the Section 106 process, into the established system of designing, building and maintaining the nation's bridges. NHPA proved to be a controversial addition to bridge management and, almost a half-century later, remains troublesome to some agencies, stakeholders, and projects.

THE SECTION 106 PROCESS

The Section 106 process is named for NHPA's Section 106, a simple one-paragraph provision requiring federal agencies to "take into account the effect of [any] undertaking on any district, site, building, structure, or object that is included in or eligible for inclusion in the National Register."[2] Section 106, says preservation law specialist John Fowler, "now has a significant impact on the way the federal government treats historic resources."[3] In its net effect, Section 106 also has a significant impact on the design, construction, rehabilitation, and maintenance of the nation's bridges, because bridge projects are often the recipients of federal-aid funds and thus invoke the Section 106 process.[4]

The Section 106 process for bridges as well as all other projects has become generally well-known and understood, if not always accepted uncritically by stakeholders. As Fowler notes, "the Section 106 process is well integrated into federal project planning," and, "in reality the Section 106 process functions to promote agreement."[5]

THE SECRETARY OF THE INTERIOR'S STANDARDS

Among the regulatory items that the Section 106 process brings to a historic bridge project are the Secretary of the Interior's Standards for the Treatment of Historic Properties. The Standards were promulgated by the National Park Service in the 1970s as part of the implementation of the federal preservation tax incentives program. In the years since, the Standards and their companion guidelines for implementation "have become the Ten Commandments for preservation work throughout the country." According to Fowler,

> They are regularly used by SHPOs and the ACHP [Advisory Council on Historic Preservation] in Section 106 cases to specify appropriate treatment for historic properties, integrated into contract specifications for federal preservation work, and used as the standard for a variety of state and local preservation laws that regulate how private property owners modify their historic properties. The Secretary's Standards are a prime example of how the federal government, without fully intending to do so and without either the desire or the ability to enforce them legally, exercises a leadership role in preservation nationwide.[6]

The Standards were assembled into their current form in 1992 and establish guidance for what should and should not be done with a historic building, structure, or landscape—including historic bridges. The Standards address four kinds of treatment: (1) preservation, (2) rehabilitation, (3) restoration, and (4) reconstruction. Each category includes a list of six to ten standards, or rules, for treatment. Most important for historic bridges are the ten standards for rehabilitation, which provide for the continuing use of structures rather than preserving them in museum-like fossilization (preservation treatment) or recreating lost structures (reconstruction treatment).[7]

HISTORIC BRIDGES, SECTION 106, AND THE SECRETARY'S STANDARDS

The years after 1966 presented an altered regulatory landscape for transportation agencies and bridge projects. Bridge engineers, long accustomed to working within standards and guidelines, now encountered a governmental element that was not necessarily in harmony with the existing process, nor was it managed by familiar agencies and professionals.

Not only was federal funding now tied to preservation issues, but there was a new activist public attitude toward the entire transportation planning process, including the preservation of historic bridges. Politically savvy citizens could use new laws and regulations to involve themselves in transportation planning and dispute agency claims about needs and priorities.[8] Transportation planning became more complex in many ways, politically as well as legally.

Bridge owners—almost exclusively government agencies—faced the possibility of the Section 106 process if a bridge was listed in, or eligible for listing in, the National Register of Historic Places. A sequence of questions emerged: Was the bridge significant and therefore eligible or not? If so, could it be replaced or not? If not, how should it be maintained? Each question could prompt significant debate.

After the wrangling over the National Register status of a bridge—a source of controversy that continues today—there remained more detailed questions of managing a historic bridge, or statewide populations of historic bridges, with established National Register listing or eligibility. Bridge owners and engineers would hear from the Federal Highway Administration (FHWA) and the State Historic Preservation Office (SHPO) that any repair and rehabilitation must meet established engineering codes and standards, but the work could not proceed without being in compliance with the Secretary of the Interior's Standards (SOIS). If the Standards were not met, the necessary federal funding could be denied.

CONVENTIONAL REHABILITATION PROCESS FOR A HISTORIC BRIDGE

Reduced to its essentials, the flowchart of a conventional rehabilitation process for a National Register listed or eligible bridge (see Figure 5.0) might include (1) plans for rehabilitation by the engineer submitted to the state department of transportation

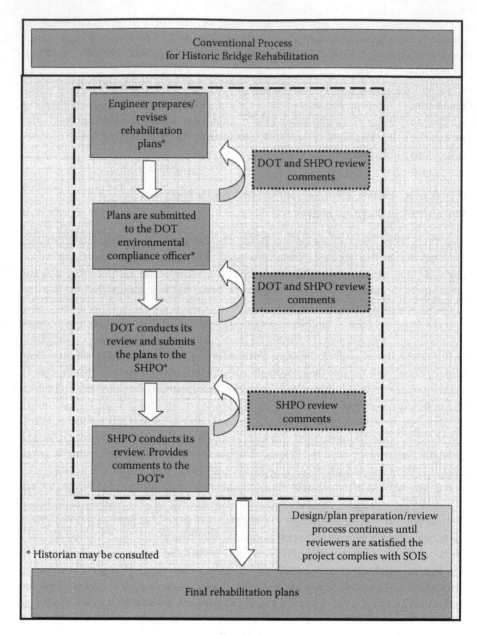

FIGURE 5.0   Flowchart for conventional process.

(DOT) environmental review and compliance officer; (2) review by DOT and sub-
mission to SHPO; (3) review by SHPO, with comment back to DOT reflecting areas
of compliance and non-compliance with SOIS; (4) return of plans to engineer with
request for revisions; (5) resubmission by engineer to DOT; continued looping of the
cycle until agreement on compliance with SOIS is achieved. The engineer, DOT,

SHPO, or all three could consult with a historian on SOIS treatment, but the nature and results would vary, depending on the historian's expertise as well as the engineer's readiness to adopt any recommendations.

This cyclical process may be brief if the proposed work involves little change to the historically significant elements of the bridge. On the other hand, the process could be very complicated if the SOIS conflict with other standards being met by the engineer, or if the engineer is unfamiliar with the process or even opposed to it. The larger and more historically significant the bridge—and the more extensive the proposed work—the more likely it is that the interaction will be complex, long, and costly. For an especially controversial project, public and political involvement may occur, and the process could escalate to the Advisory Council on Historic Preservation.

Complicating the process is the lack of familiarity, and subsequent discomfort, of some engineers with the federal preservation process in general and the SOIS in particular. Thomas F. King, an authority on federal preservation legislation and regulation, observes that "there are lots of horror stories about SHPO and NPS [National Park Service] reviewers of rehab projects getting terribly sticky about some things and being loose as geese about others, with no apparent rationale for the difference"[9]—a situation especially likely to confound engineers new to the preservation process. Unlike engineers, architects have been more familiar and comfortable with Section 106 and with SOIS, largely because of experience with rehabilitating and restoring buildings for clients taking advantage of federal preservation tax credits. King points out that "many historical architects make their living designing or consulting on the design of tax code-supported rehab projects, and others spend their lives reviewing such designs on behalf of SHPOs or NPS."[10]

How can the process be made more efficient for historic bridges? The larger process, operating at the level of stakeholders and agencies, has been studied and streamlining proposals developed. The suggested improvements involve collaborative problem-solving and mediation, parallelling other public policy debates that consume large amounts of agency time and generate political interest. For example, consider reports from the National Policy Consensus Center (NPCC) such as "Transportation Solutions: Collaborative Problem Solving for States and Communities" (2003) and its supplement, "Case Studies: Transportation Collaboration" (2003), and the report on "Transportation Collaboration in the States," developed in 2006 by NPCC for the FHWA Office of Project Development and Environmental Review.[11]

Closer to the engineering and SOIS issues involving historic bridges is the 2007 report, "Guidelines for Historic Bridge Rehabilitation and Replacement," prepared for AASHTO by Lichtenstein Consulting Engineers and Parsons Brinckerhoff Quade & Douglas.[12] While tackling the question of replacement or rehabilitation, the report touches on SOIS-related questions. In particular, the report addresses engineering issues rather than the more architecturally oriented discussions usually revolving around SOIS. This includes bridge railings, panel-point connections on trusses, arch rings and related spandrel walls, stone masonry bond patterns, decks, rivets, and geometry.

Nevertheless, the larger rehab-vs.-replace decision remains the focus rather than the engineer-historian-SOIS interaction, which is central to a plan recently developed for the Minnesota Department of Transportation.

THE MANAGEMENT PLAN FOR HISTORIC BRIDGES IN MINNESOTA: A COLLABORATIVE STRATEGY

In 2005, the Minnesota Department of Transportation (Mn/DOT) undertook an extensive historic bridge management initiative to provide comprehensive guidance for historic bridges in the state. Mn/DOT recognized that, beginning with the state's first historic bridge inventory in 1985 and continuing through subsequent surveys and inventories, the inventorying of historic bridges had largely been completed. The state's attention was turning increasingly toward systematic management of the historic bridge population. Having followed the Section 106 process on a case-by-case, bridge-by-bridge basis, Mn/DOT looked for a more efficient and comprehensive approach to the bridges it owned. At the same time, Mn/DOT wished to provide guidance to other bridge owners. After all, the state owned only 15 percent of Minnesota's National Register listed or eligible bridges. Counties and cities owned 85 percent. All owners, not only the state, struggled with the Section 106 process and all would benefit from an improved approach.

Central to Mn/DOT's initiative was the *Management Plan for Historic Bridges in Minnesota* (informally called the "general plan"), completed in 2006 by consulting firms Mead & Hunt and HNTB.[13] The two firms were selected because of their extensive experience with historic bridges—Mead & Hunt with historic bridge survey, inventory, and the Section 106 process, and HNTB with the repair and rehabilitation of historic bridge structures. Pairing the two firms in the preparation of the plan would also model the process that Mn/DOT envisioned at the heart of its initiative, a collaborative relationship of history and engineering, and historian and engineer. At the same time, Mn/DOT would take the process one giant step beyond the general plan by having the same consulting team complete individual management plans for 22 state-owned historic bridges. Each individual plan would employ the collaborative team process outlined in the general plan.

The general management plan lays out the goals, strategies, and process for completing the individual management plans. It also includes supplementary information for historic bridge management, such as technical guidance on recommended stabilization, preservation, maintenance, and inspections, as well as special considerations for relocating bridges and for the application of design exceptions and variances. The general plan includes a discussion on options for funding historic bridge projects. Additional information in the appendix identifies important contacts for additional assistance, a list of National Register listed and eligible bridges in the state, and a sample individual management plan for a Minnesota historic bridge.

The general plan's heart and procedural focus is the collaborative process for developing the individual plans. This process begins with an overarching preservation goal: "to preserve a historic bridge in the way that best retains the qualities that give it historic significance while meeting transportation needs."[14] Mn/DOT identified five options for meeting the goal, one of which is to be selected for each individual plan. In order of descending preference, the options are as follows:[15]

Rehabilitation for continued vehicular use on-site. This is the preferred option because it represents the best combination of retaining historical features while meeting transportation needs.

Rehabilitation for less-demanding use on-site. The bridge remains in use on-site as part of a one-way pair or other reduced use.

Relocation and rehabilitation for less-demanding use. The bridge is relocated to another site, such as a pedestrian and bicycle trail.

Closure and stabilization, pending future use. This is a variation of option 3 (relocation), but is implemented when a relocation alternative is not yet available.

Partial reconstruction while preserving substantial historic fabric. This is the least-desirable option because it results in the greatest loss of historic features.

Each individual plan will specify the recommended preservation option as well as detailed recommendations (with cost estimates) for preservation, rehabilitation, and maintenance. All of this—the survey, evaluation, preparation of recommendations, and identification of preferred preservation option—is established through the collaborative team approach. The team approach is summarized in the general plan as follows:

> In the team approach, a professional historian and a professional engineer, both experienced in historic bridge evaluation, conduct the field surveys together. Following field survey, they review issues and discuss stabilization, preservation, and maintenance recommendations together for each bridge. A management plan for each bridge is prepared that includes the analyses and recommendations of this interactive approach.[16]

The general plan lays out the roles for the historian and the engineer (see Figure 5.1). The historian (1) establishes the individual bridge's significance as presented in the National Register documentation, (2) identifies character-defining features, and (3) applies the SOIS in collaboration with the project engineer. The general plan states: "the Secretary's Standards are central to the dialogue between the historian and the engineer. Use of these standards can connect the requirements of historic preservation laws with transportation needs and guide the engineer's recommendations for bridge stabilization, preservation, and maintenance."[17]

Meanwhile, the engineer (1) assesses transportation needs at the site, (2) assesses the condition of the bridge, (3) assesses rehabilitation needs, and (4) estimates costs.[18]

In practical application, the historian and engineer on a bridge project will work closely, "in the same timeframe," and "tightly integrated into a common schedule." The process begins with conducting the bridge survey together, each noting the details and features that the other observes. As the plan states, "Each professional evaluates the bridge with the other's interests, concerns, and regulations in mind. They are interacting regularly with the understanding that any final recommendations must accomplish the combined purpose of historic bridge management." The dialogue is especially important where the SOIS are involved. Here, the discussion is in-person and face-to-face as much as possible. In the end, the intensive team

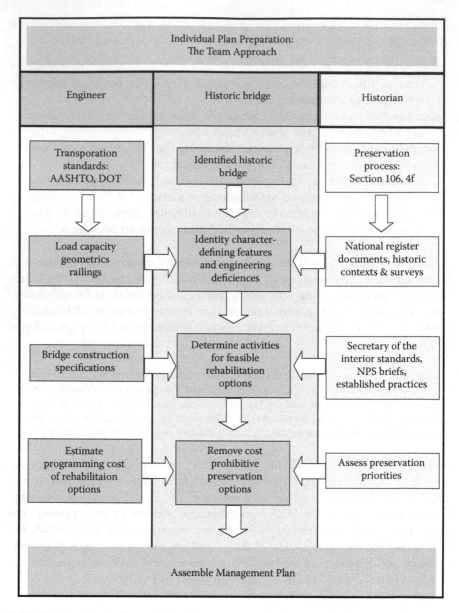

FIGURE 5.1 Flowchart for collaborative team process.

approach "provides predictability for the bridge owner, who can have a better understanding of the regulatory outcome, earlier in the process, for the historic bridge."[19]

APPLYING THE COLLABORATIVE APPROACH TO
INDIVIDUAL BRIDGES: MINNESOTA EXAMPLES

The same consulting team of a professional historian from Mead & Hunt and a professional engineer from HNTB produced individual management plans for 22 state-owned

historic bridges to complement the general management plan. The 22 bridges had been selected by Mn/DOT as candidates for long-term preservation and therefore would require management plans.[20] Each bridge was listed in the National Register, but the group represented a considerable range of conditions to be considered in management plans. Each individual plan employed the collaborative team process described in the general plan.

In this project, the execution of the collaborative approach to historic bridge management was as varied as the bridges being evaluated. Bridges in good condition, with adequate load capacity, geometrics and unaltered character-defining features, presented uncomplicated cases for the historian and engineer to apply the SOIS. As the team evaluated bridges with successively deteriorated structural conditions, or as the level of rehabilitation necessary to keep the structure on the highway system increased, the amount of discussion between the engineer and historian increased. As the level of needed rehabilitation increased, the effect on character-defining features and historic fabric was also likely to increase.

Examples of the dialogue between historian and engineer employing the SOIS are reviewed below. In each case, the historian and the engineer visited the bridge together. While the engineer assessed the conditions in relation to available inspection reports, the historian reviewed the structure in relationship to the National Register documentation. Returning to their offices, the engineer prepared preliminary proposals for preservation, rehabilitation, and maintenance, while the historian developed a list of character-defining features. The two subsequently met and reviewed the engineer's proposals, the character-defining features, and the implications of each. Through in-person dialogues, the recommended preservation option was determined, followed by a revised set of proposals that would meet the SOIS.

BRIDGES REQUIRING MINOR REHABILITATION WITH ADEQUATE GEOMETRICS AND LOAD CAPACITY

Historic bridges in good condition with good geometrics and load capacity were the easiest subjects for management plan development. Extensive rehabilitation was not necessary for these bridges to stay on Minnesota's trunk highway system. The management plans for these bridges focused primarily on preservation and maintenance activities. The character-defining features were not affected by work needed for the bridge to continue in service.

Bridge 4700 (Sorlie Memorial Bridge, built 1929, Figure 5.2 and Figure 5.3) is an example of a bridge requiring minimal rehabilitation.[21] This two-span Parker truss bridge carries U.S. 2 over the Red River of the North, connecting the communities of East Grand Forks, Minnesota, with Grand Forks, North Dakota. The structural condition codes for the bridge varied from fair to good with the need for repainting the bridge topping the list of work items. No work was recommended that would affect the character-defining features (original truss design, ornamental railing, and truss-support trucks), or other historic fabric. Compliance with SOIS was not difficult.

FIGURES 5.2 AND 5.3   Bridge 4700. *Top*: Elevation view. *Bottom*: Truss-support trucks, a character-defining feature.

Bridges Requiring Minor Rehabilitation with Adequate Geometrics and Unknown Load Capacity

Bridges in fair or better condition with good geometrics and unknown load carrying capacity were another group with short lists of required stabilization activities. These were relatively small bridges on the trunk highway system with steel Multi Plate arches that had never been formally load rated. As constructed in Minnesota, steel Multi Plate arches are assembled using plates of curved, corrugated, galvanized steel. The segments are bolted together in the field to form a semi-circular shape that is supported on masonry or concrete elements. The Multi Plate field-bolted arch was introduced in 1931 by the Armco Culvert Manufacturers Association.[22] Headwalls are constructed at each end of the arch to retain the fill placed over the crown of the arch to support the roadway. Originally classified as culvert structures, their load capacity was simply obtained from a standard chart identifying load capacity for different types of culverts.

For these Multi Plate arches, it was recommended that a more rigorous load rating analysis be performed with soil-structure interaction to ensure that there was adequate fill over the bridge. If the resulting load capacity was found to be insufficient, it was recommended that a distribution slab be integrated into the pavement above the structure. One example is Bridge 5827, a WPA structure built in 1938 using Multi Plate (Figure 5.4).[23] The character-defining features for this bridge included the Multi Plate metal arch and the stone masonry head- and wing-walls. Stabilization activities were recommended to preserve and protect these features of the bridge. The work items recommended in the management plan resulted from an uncomplicated dialogue between historian and engineer using the SOIS.

Bridges Requiring Minor Rehabilitation with Adequate Geometrics and Poor Load Capacity

Bridge 3355, built in 1921 and widened in 1939 with stone-masonry treatment, is a small reinforced-concrete slab bridge that required more dialogue between historian and engineer (Figure 5.5).[24] The conversation focused on how to improve the load capacity to keep the bridge in service on the trunk highway system, while meeting SOIS. The bridge was designed to carry H10 loading and, when load rated analytically, was found to have deficient capacity. The inventory load rating was HS12 and the operating rating was HS18 (based on a physical inspection rating). While currently functioning as originally intended, discussion indicated that continuing vehicular use on-site was not a recommended preservation option that would be satisfactory for the 20-year life of the management plan. The bridge would begin to deteriorate as truck loads continued to increase.

The span of Bridge 3355 was quite short and the fill over the slab was several feet. The character-defining feature for this bridge was not the reinforced-concrete slab, but the granite masonry used for the headwalls and wingwalls. SOIS compliance directed the team to retain as much historic fabric as possible, including the original slab superstructure if possible. The consensus reached by the engineer and historian was to embed a new structural slab over the existing bridge. This

FIGURE 5.4   Bridge 5827, a large Multi Plate arch bridge with WPA stone-masonry head- and wing-walls.

recommendation had several benefits. It would retain both the original slab and the character-defining masonry while providing load capacity meeting current standards as well as a structural element to anchor new traffic railings inside the existing stone masonry railings.

BRIDGES REQUIRING MAJOR REHABILITATION WITH POOR GEOMETRICS AND POOR LOAD CAPACITY

Bridges with marginal load capacity and geometrics were among the most difficult to work through. Recommendations to improve the geometrics to provide adequate lane widths and shoulder widths affected character-defining features beyond the degree that would ensure compliance.

Bridge 5721 is an example of this difficult combination of elements (Figure 5.6 and Figure 5.7).[25] It is a circa 1870's iron through truss with deficient geometrics, marginal load capacity, and a deteriorating floor system and approach spans. Originally erected elsewhere in Minnesota, it was relocated to its current site in 1937.

FIGURE 5.5   Bridge 3355.

The plans document its fabrication in iron. With several major deficiencies, each improvement necessary to update it to meet current standards presented a major negative impact to its character-defining features. Widening the truss to meet lane and shoulder width standards, would mean modifying or replacing all of the floor beams and the top and bottom bracing. In addition, the extra dead and live load associated with the wider bridge would exacerbate the load capacity problem. Providing a parallel structure and reconfiguring the historic bridge as part of a one-way pair of bridges could improve the geometric system, but would not improve the load capacity of the historic bridge. In addition, the financial and logistical feasibility of providing a second bridge at the site was deemed unlikely.

Because truss bridges are much easier to relocate than others, moving this bridge off-system would be the recommended preservation option. The fact that the truss had been relocated to its current site within the historic period provided support for the recommendation. Currently this truss is scheduled for relocation to a state department of natural resources trail.

OFF-SYSTEM BRIDGES

Two of the bridges evaluated for individual management plans are off-system bridges owned by Mn/DOT: a major stone arch bridge, built as a railroad bridge but converted to a pedestrian/bicycle bridge; and a large deck-truss structure. Because of their off-system status, the preservation options are different than bridges currently in highway use.

FIGURES 5.6 AND 5.7 Bridge 5721. *Top*: Elevation View. *Bottom*: View from South Approach.

FIGURE 5.8   Northwest face of Stone Arch Bridge (Bridge 27004), showing six-degree curve.

The James J. Hill Stone Arch Bridge (Bridge 27004, built in 1881–1883) crosses the Mississippi River on a six-degree curve in downtown Minneapolis (Figure 5.8).[26] Situated within a National Register historic district, the 2,100-foot bridge is also individually listed on the National Register and is an ASCE National Civil Engineering Landmark. A previous project to convert its use from the original railroad function to bicycle/pedestrian use had necessitated substantial changes determined to be in compliance with SOIS.

There was no question that this structure would have a management plan recommending preservation on-site and continued use as a pedestrian facility. The character-defining features include the horizontally curved alignment and the stone masonry construction. The management plan proposals involved largely maintenance issues of water infiltration, which required no SOIS negotiation.

A more difficult off-system problem was posed by Bridge 4715 in Shakopee, Minnesota, a large deck-truss bridge built in 1927 over the Minnesota River (Figure 5.9).[27] With its superstructure and substructure in poor condition and functioning as an ad hoc pedestrian facility by virtue of being closed to vehicular traffic, it was a challenge to the engineer. The deck-truss superstructure was identified as a character-defining feature. Preserving the truss to function as an extremely wide pedestrian facility was awkward. A much narrower bridge could serve as a pedestrian link over the Minnesota River. Preserving and rehabilitating two of the three truss lines as part of a narrower bridge was dismissed due to the negative impact to one of the

FIGURE 5.9   Bridge 4715.

primary character-defining features (superstructure). It was an expensive option for the team to recommend rehabilitation of the complete superstructure compared to the costs of a new (much narrower) pedestrian bridge. In the end, the team concurred with Mn/DOT that the 22 bridges selected for long-term preservation were part of the greater collection of bridges owned by Minnesota and that the large costs to rehabilitate this bridge should be considered from a system perspective and not an individual project perspective.

Park Bridges

The collaborative historian/engineer approach to bridge management has also been successfully used for two park bridges owned by St. Paul, Minnesota. These two separate projects illustrate how the collaboration adapts to the case at hand.

One example is L-5853, a 1904 Melan-system concrete arch bridge, now the second-oldest reinforced concrete arch extant in Minnesota (Figure 5.10).[28] It was built as a pedestrian bridge over streetcar tracks in Como Park by the streetcar company to allow riders to safely cross the tracks and reach the adjacent streetcar station, which still exists. The final recommendations for this bridge were largely driven by the bridge's historical significance and its location in a city park. With the streetcar long gone, there is no functional need for the bridge. It is in poor condition. The city, however, preferred retaining the historic park structure if it could be preserved economically and would allow a trail to pass underneath on the old grade. Park officials were responsive to a management-plan proposal to retain the significant concrete arch while exposing elements of the Melan truss in situ, allowing public viewing with interpretive signage (Figure 5.11). The center deck section over the arch would be removed, with remaining deck elements preserved as viewing platforms to observe the exposed Melan system elements. This compromise solution retains the character-defining feature-the Melan arch-at a cost far less than reconstructing the entire bridge and, in the process, concealing the Melan artifacts. In fact, it was the cost of reconstructing an "unnecessary" functional historic bridge that accelerated city interest in demolition over preservation. The exposed arch structure would now be considered an educational and historical amenity to the park that also had elements of public art. The project currently awaits funding, but the prospects are very positive.

The second example is Bridge L-8560, a stone-veneered, reinforced concrete arch in St. Paul's Phalen Park, built in 1910 with a design by C.A.P. Turner and rebuilt with a stone-masonry veneer in 1934 (Figure 5.12).[29] In contrast to the Como Park bridge, the recommendations for the Phalen Park bridge were driven largely by engineering considerations for rehabilitation. Unlike Como, the Phalen bridge remains in use as a pedestrian bridge and continued use is desired by the city. A citizens group supports rehabilitation of this deteriorated bridge rather than replacement with a modern pedestrian structure. The collaborative recommendation involves the use of innovative engineering elements of micropiles and precast concrete liners to meet SOIS and retain its character-defining features of a barrel arch and masonry components (Figure 5.13).

SUMMARY AND CONCLUSION

The collaborative team approach to historic bridge management was presented in detail in the *Management Plan for Historic Bridges in Minnesota* and adopted by Mn/DOT for its larger historic bridge initiative. At the same time, the approach was employed in 22 individual management plans for state-owned historic bridges of varying types and conditions. Mn/DOT's intention was to provide a more efficient process for managing a large, statewide population of historic bridges through a collaborative team approach. The team approach paired a professional engineer with a professional historian in the actual plan development rather than trying to combine two separate evaluations at a later date. Such a collaborative process would reduce the typical negotiation cycle involving the SOIS, thus saving time and money. At the same time, the process would give Mn/DOT, the bridge owner, a recommended preservation option and a comprehensive overview of the long-term resources required for the implementation of the option.

The approach has proved workable in all cases, including the 22 state management plans, and the plans for two city park bridges. The historian-engineer team effectively combined the observations from the two professional areas and resolved bridge preservation issues involving the SOIS, regardless of the complexity of the problems encountered in each bridge. Producing management plans for all the bridges in a single project allowed the team to develop experience with a range of historic bridge problems and with the collaborative approach to problem-solving. The collective Mn/DOT project thus embodied continuity of approach and process across the 22 bridge plans. The work was completed to the satisfaction of both a state bridge owner and a city bridge owner.

FIGURES 5.10 AND 5.11   Bridge L-5853, Como Park. *Top*: View from below. *Bottom*: Preservation concept.

FIGURE 5.12   Phalen Park bridge (L-8560) arch soffit.

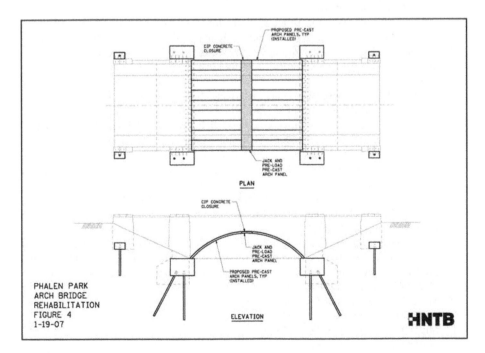

FIGURE 5.13   Phalen Park bridge (L-8560) rehabilitation strategy.

Based on this extended experience, the collaborative process as presented here is worthy of emulation by other agencies needing to develop historic bridge rehabilitation and management plans with minimal conflict and delay and potentially a savings in project time and expense.

REFERENCES

1. Diana Lea, "America's Preservation Ethos: A Tribute to Enduring Ideals," in *A Richer Heritage: Historic Preservation in the Twenty-First Century*, Robert E. Stipe, ed. (Chapel Hill: University of North Carolina Press, 2003), 11–12.
2. *Federal Historic Preservation Laws* (Washington, D.C.: National Park Service, 2002), 59.
3. John M. Fowler, "The Federal Preservation Program," in *A Richer Heritage*, 45.
4. A parallel, but more stringent, preservation process results from the invocation of Section 4(f) of the Department of Transportation (DOT) Act of 1966. For the purposes of this paper, the two processes can be considered similar enough to be discussed generally as Section 106. See Fowler, 52–54.
5. Fowler, 50.
6. Fowler, 62.
7. Kay D. Weeks and Anne E. Grimmer, *The Secretary of the Interior's Standards for the Treatment of Historic Properties* (Washington, D.C.: National Park Service, 1995). As noted in the flyleaf to this volume, the first Standards were codified in 1978 and 1983 as 36 CFR 68, "The Secretary of the Interior's Standards for Historic Preservation Projects," and revised in 1992 and codified as 36 CFR 68 in the July 12, 1995 Federal Register (Vol. 60, No. 133), as the Secretary of the Interior's Standards for the Treatment of Historic Properties."
8. See extensive discussion in Bruce E. Seely, *Building the American Highway System: Engineers as Policy Makers* (Philadelphia: Temple University Press, 1987), 225–237.
9. Thomas F. King, *Cultural Resource Laws and Practice: An Introductory Guide*, 2nd edition (Walnut Creek, Calif.: AltaMira Press, 2004), 237.
10. King, 245.
11. National Policy Consensus Center, Portland State University, Portland, Oregon, "Transportation Solutions: Collaborative Problem Solving for States and Communities" (2003), "Case Studies: Transportation Collaboration" (2003), and "Transportation Collaboration in the States" (2006), all available at *www.policyconsensus.org*.
12. "Guidelines for Historic Bridge Rehabilitation and Replacement," prepared by J. Patrick Harshbarger, Mary E. McCahon, Joseph J. Pullaro, Steven A. Shaup of Lichtenstein Consulting Engineers in association with Parsons Brinckerhoff Quade & Douglas, for American Association of State Highway and Transportation Officials (AASHTO), as part of NCHRP Project 25-25/ Task 19, National Cooperative Highway Research Program, Transportation Research Board, March 2007.
13. *Management Plan for Historic Bridges in Minnesota*, prepared by Mead & Hunt and HNTB for Minnesota Department of Transportation, June 2006. Copy in PDF format available through link at www.dot.state.mn.us/environment/cultural_res/index.html
14. Mn/DOT, *Management Plan*, 33.
15. Mn/DOT, *Management Plan*, 33–35.
16. Mn/DOT, *Management Plan*, 36.
17. Mn/DOT, *Management Plan*, 37–39.
18. Mn/DOT, *Management Plan*, 40–42.
19. Mn/DOT, Management Plan, 37.

20. See "Foreword: Managing Minnesota's historic bridges," Mn/DOT, *Management Plan*, F-i-ii.
21. Minnesota Department of Transportation, Historic Bridge Management Plan for Bridge 4700, June 2006. Copies of this and other individual management plans available at Mn/DOT, Office of Environmental Services, Cultural Resources Unit, MS 620, 395 John Ireland Blvd, St. Paul, Minnesota, 55155.
22. See discussion of the Multi Plate Arch in "Iron and Steel Bridges in Minnesota," National Register of Historic Places, Multiple Property Documentation Form, prepared July 1988, listed September 29, 1989, E19-20, F10.; copy available in Minnesota State Historic Preservation Office. See also Armco Culvert Manufacturers Association, *Handbook of Culvert and Drainage Practice*, 2nd edition (Chicago, 1937), 58.
23. Minnesota Department of Transportation, Historic Bridge Management Plan for Bridge 5827, June 2006.
24. Minnesota Department of Transportation, Historic Bridge Management Plan for Bridge 3355, June 2006.
25. Minnesota Department of Transportation, Historic Bridge Management Plan for Bridge 5721, June 2006.
26. Minnesota Department of Transportation, Historic Bridge Management Plan for Bridge 27004, June 2006.
27. Minnesota Department of Transportation, Historic Bridge Management Plan for Bridge 4715, June 2006.
28. "Bridge L-5853, Lexington Avenue Pedestrian Bridge in Como Park: Preservation Option Study," prepared by HNTB and Mead & Hunt for City of St. Paul, Division of Parks and Recreation, June 24, 2007. Copy available from St. Paul Division of Parks and Recreation, 300 City Hall Annex, 25 West Fourth St., St. Paul MN 55102.
29. "Phalen Park Bridge L-8560: Engineering Assessment and Historic Evaluation," prepared by HNTB and Mead & Hunt for City of St. Paul, Division of Parks and Recreation, January 23, 2007. Copy available from St. Paul Division of Parks and Recreation, 300 City Hall Annex, 25 West Fourth St., St. Paul MN 55102.

Part 3

Evaluation

6 Structural Deck Evaluation of the John A. Roebling Suspension Bridge

Ching Chiaw Choo and Issam E. Harik

CONTENTS

INTRODUCTION

THE JOHN A. ROEBLING BRIDGE

Completed in 1867, the John A. Roebling Bridge (see Figure 6.1)—formerly the Covington-Cincinnati Suspension Bridge—was the first permanent bridge to span the Ohio River between Kentucky and Ohio. In 1975, the bridge was designated as a National Historic Civil Engineering Landmark by the American Society of Civil Engineers and was listed on the National Register of Historic Places.

The John A. Roebling Bridge, which carries KY 17 over the Ohio River between the two aforementioned cities, is a three-span bridge. The main span of the bridge is approximately 1,100-ft long, carrying a two-lane 28-ft wide roadway. The two approach spans are approximately 300-ft long; resulting in a superstructure of approximately 1700-ft long. The bridge also carries an 8-ft 6-inch wide sidewalk cantilevered from both sides of the superstructure. The roadway's structure consists of a steel grid decking system, structural channel (**C**) sections, structural standard (**S**) sections, and built-up I-shaped plate girders. The roadway structural system is subsequently supported by planar trusses, secondary suspenders, and primary cables. In 2007, the bridge's weight restrictions are posted as 17 tons for two-axle trucks and 22 tons for three-, four-, and five-axle trucks. Numerous structural truss and floor system repairs had been made in the past, with the latest one in the early 1990s.

RESEARCH OBJECTIVE AND SCOPE

The objective of this study is to conduct a structural evaluation of the John A. Roebling Bridge in order to determine the maximum allowable gross vehicle weight (GVW) that can be carried by the roadway or bridge deck structural elements shown in Figure 6.2 (i.e., open steel grid decking, channel sections, standard sections, and built-up sections).

CAPACITY EVALUATION OF THE BRIDGE DECK ELEMENTS

The bridge deck consists of the following four (4) structural components: open steel grid decking system (Figure 6.3), structural channel (**C**) section (Figure 6.4), structural standard (**S**) section (Figure 6.5), and built-up I-shaped plate girder (Figure 6.6).

FIGURE 6.1 The John A. Roebling Bridge carries KY 17 over the Ohio River between Covington, KY, and Cincinnati, OH.

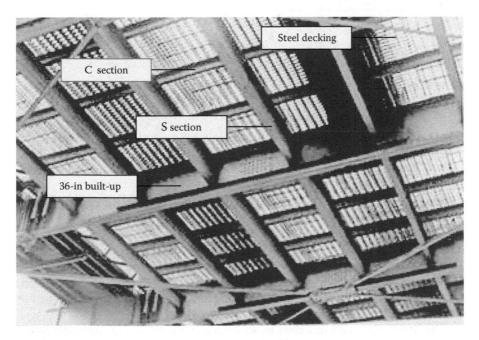

FIGURE 6.2 Structural elements of the John A. Roebling Bridge.

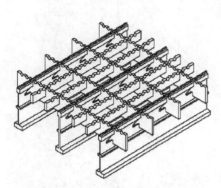

FIGURE 6.3 5-inch steel deck.

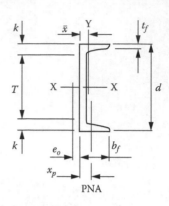

FIGURE 6.4 Channel (C) section.

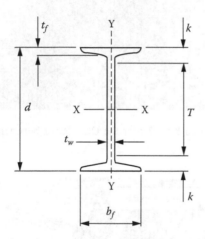

FIGURE 6.5 Standard (S) section.

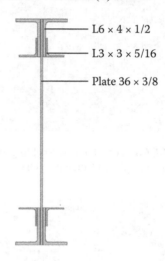

FIGURE 6.6 36-inch built-up section.

Steel Decking

Due to a lack of details, the capacity of the existing open steel decking was determined by comparing the existing deck to a similar type of commercially available
steel decking. The dimensions of the existing steel decking were measured to be
5-1/4" in height with main rails spaced 6-in center-to-center.

One commercially available open steel grid decking manufactured by the Interlocking Deck Systems International (IDSI), Inc.,[1] was found to be comparable to
the existing steel decking. The main rail of the IDSI's steel decking has a height
of 5-3/16", as shown in Figure 6.7. The complete steel grid decking system (see
Figure 6.8) consists of cross bars in the two perpendicular directions, main rails in
the traffic direction, and reinforcing bars (not shown) in the transverse direction.
Section properties of the IDSI's steel decking with mail rails spaced at 6-in center-to-
center are presented in Figure 6.9.

The IDSI's decks are designed in accordance with AASHTO Allowable Stress
Design.[2] The load capacities of steel decks with 6-in main rail spacing are tabulated

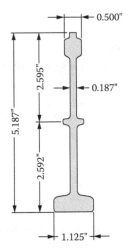

FIGURE 6.7   Main rail of the IDSI deck.[1]

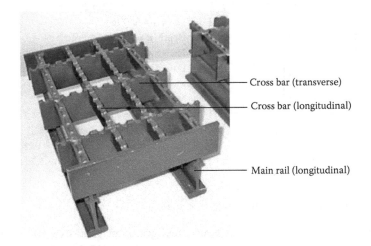

FIGURE 6.8   Components of the IDSI deck.

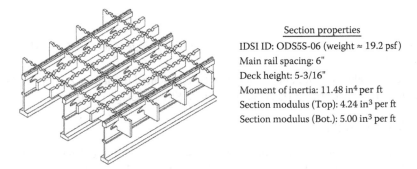

Section properties
IDSI ID: ODS5S-06 (weight ≈ 19.2 psf)
Main rail spacing: 6"
Deck height: 5-3/16"
Moment of inertia: 11.48 in^4 per ft
Section modulus (Top): 4.24 in^3 per ft
Section modulus (Bot.): 5.00 in^3 per ft

FIGURE 6.9   Section properties of the IDSI's deck with 6-in main rail spacing.[1]

TABLE 6.1
Load Table for IDSI Steel Decks[1]

	HS 20 (MS 18) Max. Cont. Clear Span			HS 25 (MS 22) Max. Cont. Clear Span		
	Transverse/Parallel to Traffic		Deflection	Transverse/Parallel to Traffic		Deflection
IDSI ID	36 ksi	50 ksi	L/800	36 ksi	50 ksi	L/800
ODS5S-06	5.34 ft	7.19 ft	5.61 ft	5.01 ft	6.74 ft	5.43 ft

Notes:
- Apply only for ODS5S-06 where mail rails are 6-in center-to-center.
- Modulus of elasticity of steel decks = 29×10^6 psi.
- Clear span = L.
- Deflection limits shown are independent of the main rail orientation for AASHTO ASD method.[2]
- Steel strength limits = 27 ksi for 50 ksi yield steel or 20 ksi for 36 ksi yield steel.
- Fatigue was not considered.

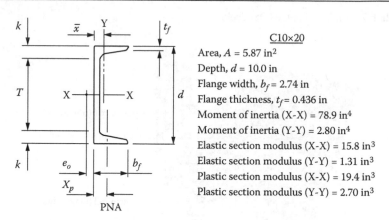

C10×20
Area, A = 5.87 in²
Depth, d = 10.0 in
Flange width, b_f = 2.74 in
Flange thickness, t_f = 0.436 in
Moment of inertia (X-X) = 78.9 in⁴
Moment of inertia (Y-Y) = 2.80 in⁴
Elastic section modulus (X-X) = 15.8 in³
Elastic section modulus (Y-Y) = 1.31 in³
Plastic section modulus (X-X) = 19.4 in³
Plastic section modulus (Y-Y) = 2.70 in³

FIGURE 6.10 Properties of C10×20 channel section.

in Table 6.1.[1] As illustrated, two different steel grades of yield strengths 36 ksi and 50 ksi, respectively, are considered for design trucks of HS 20 (MS 18) and HS25 (MS 22)—25% weight or load increase of the HS 20 truck type.[2]

Steel Channels (C)

The steel channel (C) used in the John A. Roebling Bridge is a C10×20, as shown in Figure 6.10.

The allowable flexural and shear capacities presented in Table 6.2 of the C10×20 were determined per the 2005 AISC Allowable Stress Design (ASD).[3]

Steel Standards (S)

There are two types of steel standards (S) used in the John A. Roebling Bridge: S15×50 and S20×66. Only the capacities of the S15×50 (Figure 6.11) were evaluated as it is the smaller and the more critical structural member in this case.

TABLE 6.2
Flexural and Shear Capacity of the C10×20 Channel Section

Member	Steel Grade	Allowable M_a	Allowable V_a
C10×20	A36 (36 ksi)	31.3 k-ft	49.0 k

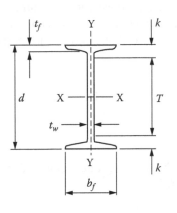

S15×50
Area, A = 14.7 in²
Depth, d = 15.0 in
Flange width, b_f = 5.64 in
Flange thickness, t_f = 0.622 in
Web thickness, t_w = 0.550 in
Moment of inertia (X-X) = 485 in⁴
Moment of inertia (Y-Y) = 15.6 in⁴
Elastic section modulus (X-X) = 64.7 in³
Elastic section modulus (Y-Y) = 5.53 in³
Plastic section modulus (X-X) = 77.0 in³
Plastic section modulus (Y-Y) = 9.99 in³

FIGURE 6.11 Properties of S15x50 standard section.

TABLE 6.3
Flexural and Shear Capacity of the S15×50 Standard Section

Member	Steel Grade	Allowable M_a	Allowable V_a
S15×50	A7 (33 ksi)	95.0 k-ft	109.0 k

TABLE 6.4
Flexural and Shear Capacity of 36-Inch Built-Up Section

Member	Steel Grade	Allowable M_a	Allowable V_a
36-inch built-up	A36 (36 ksi)	549.7 k-ft	175.0 k

The allowable flexural and shear capacities presented in Table 6.3 of the standard section S15×50 were determined per the 2005 AISC Allowable Stress Design (ASD).[3]

STEEL BUILT-UP SECTIONS

The built-up members supporting the open grid steel decking, channels, and standards, have a 36-in height and are composed of four angles L6×4×(1/2) (two at top and two at bottom), four angles L3×3×(5/16) (two at top and two at bottom), and a steel plate 36×(3/8), as shown in Figure 6.6. Intermediate web stiffeners of the built-up sections are not shown in the figure.

The allowable flexural and shear capacities presented in Table 6.4 for the built-up members were determined per the 2005 AISC Allowable Stress Design (ASD).[3]

BRIDGE LOADINGS

Two types of gravity loads are considered in the analysis: self-weight loads of the structural and non-structural elements, and live loads.

SELF-WEIGHT LOADS

Attributable self-weight loads include loads of the bridge deck structural elements (i.e., steel decking, channel sections, standard sections, and built-up sections) and non-structural elements (i.e., electrical and mechanical conduits, traffic signs, posts, etc.).

LIVE LOADS

By definition, live loads are transient loads. In this study, live loads are contributed by the different truck and bus types.

Truck Types

The four truck types, commonly considered by the state transportation department, traversing the bridge are presented in Table 6.5.

Bus Types

The three bus types, provided by the local transit, traversing the bridge are presented in Table 6.6.

ASSUMPTIONS

The following assumptions are introduced in the analysis:

- A 30% impact load is considered in the analysis. This is in accordance with the 2002 AASHTO Standard Specification Section 3.8.2.[2]
- Two vehicles (i.e., trucks and/or buses) can travel parallel to each other on the bridge at the same time to produce the maximum load effect. This condition applies to certain structural elements (i.e., channel sections and the 36-in built-up member).

ELEMENT SECTIONAL AND STRENGTH CAPACITY LOSSES

Recent field inspections revealed that some structural elements in the deck have experienced sectional loss up to 20%. The loss can be attributed to rust, visible cracks, etc. An accurate estimate of the section loss requires element removal from the bridge, cleaning, detailed measurements, etc. Consequently, an estimate based on visual inspection and field measurements is more practical. However, only visible losses can be measured, and these generally underestimate the actual section losses (e.g., cracks that are not visible to the naked eye, etc).

The sectional losses reduce the sectional geometric properties of the element (i.e., area A, moment of inertia I, section modulus S, etc.) and, in turn, reduce the strength capacity of the section in bending, shear, etc.

TABLE 6.5
Trucks Traversing the Roebling Bridge

Truck Type	Truck Information	
	Axle Spacing s	Wheel Spacing s_w
Type 1 s, s_w 0.2 W, 0.8 W	s = 14'-0"	6'-0"
Type 2 s_1, s_2, s_w 0.14 W, 0.43 W, 0.43 W	s_1 = 12'-0" s_2 = 4'-0"	6'-0"
Type 3 s_1, s_2, s_2, s_w 0.19 W, 0.27 W, 0.27 W, 0.27 W	s_1 = 12'-0" s_2 = 4'-0"	6'-0"
Type 4 s_1, s_2, s_3, s_2, s_w 0.12 W, 0.22 W, 0.22 W, 0.22 W, 0.22 W	s_1 = 12'-0" s_2 = 4'-0" s_3 = 14'-0"	6'-0"

In order to quantify the relation between the percentage of section loss and the percentage of capacity loss, results are presented in tables in section 5 for 10% to 40% loss in section, in 10% increments. The percentage loss is applied uniformly to the flanges and webs of the steel sections (e.g., **C** and **S** sections) and to the walls of the steel sections that make up the built-up member. For example, a 10% section loss is applied by reducing the thickness of the flanges and webs by 10%.

Table 6.7 shows that, a 10% section loss leads to a 19% loss in allowable bending moment capacity and 10% loss in allowable shear capacity of the built-up section. A 20% section loss leads to 38% loss in allowable bending moment capacity and 20% loss in allowable shear capacity of the built-up section.

ELEMENT LEVEL ANALYSIS

The *Element Level Analysis* is a level of analysis that treats the different structural components as individual components. In general, this level of analysis yields a

TABLE 6.6
Buses Traversing the Roebling Bridge

Bus Type	Bus Information		
	Gross Vehicle Weight* W	Axle Spacing** s	Wheel Spacing** s_w
Type 1 29-ft	30,000 lbs (15.00 Tons)	13' 6"	8' 1" (+/– 1")
Type 2 35-ft	39,500 lbs (19.75 Tons)	18' 4"	8' 3" (+/– 1")
Type 3 40-ft	39,500 lbs (19.75 Tons)	23' 8"	8' 3" (+/– 1")

* The Gross vehicle weight is the weight for the fully loaded bus. The information was provided by the Transit Authority of Northern Kentucky (TANK).

** *Information provided by the Transit Authority of Northern Kentucky (TANK).*

conservative estimate of capacity. Each element or component is assigned a specific tributary area, and it is assumed that there is no load contribution and/or distribution to and/or from adjacent members. Support conditions are idealized as appropriate (i.e., simple, fixed, etc).

STRUCTURAL IDEALIZATION

Steel Decking

The following assumptions are applied to the open steel grid decking for this level of analysis:

- The deck is to be continuously supported over several spans; and
- The channels (**C**), supporting the deck, are idealized as simple supports (Figure 6.12).

TABLE 6.7

Effect of 10%, 20%, 30%, and 40% Sectional Loss on the Sectional Properties and Capacities of the 36″ Built-Up Member

Section Properties and Allowable Shear and Bending Capacities	0% Sectional Loss	Value of the Section Properties and Capacities for Different % in Sectional Loss							
		10% Sectional Loss		20% Sectional Loss		30% Sectional Loss		40% Sectional Loss	
	Value	Value	% Reduction = η	Value	% Reduction = η	Value	% Reduction = η	Value	% Reduction = η
Area, A (in²)	39.62	32.90	17%	26.79	32%	22.18	44%	17.41	56%
Moment of inertia, I_x (in⁴)	7,580	5,923	22%	4,750	37%	3,686	51%	2,702	64%
Elastic section modulus, S_x (in³)	421	340	19%	260	38%	204	52%	150	64%
Plastic section modulus, Z_x (in³)	523	422	19%	322	38%	251	52%	188	64%
Shear capacity, V_a (k)	175	157	10%	139	20%	121	31%	104	41%
Bending moment capacity, M_a (k-ft)	549.7	445	19%	341	38%	264	52%	198	64%

Notes:

• The sectional loss in the bridge elements may occur as a result of a crack propagating in the web or the flange(s). In this case, the section properties and capacities listed in column 1 in Table 6.7 can be derived based on the uncracked section in order to determine % reduction.

• In case a % sectional loss falls between two values in Table 6.7 (e.g. 14% sectional loss), a linear interpolation between the % sectional loss that is lower and the one that is higher than the one in question (e.g. 10% and 20% sectional loss) should yield adequate results.

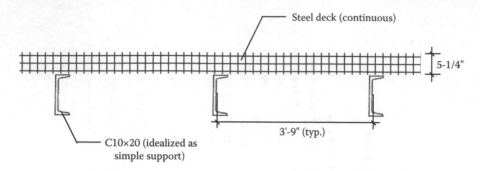

FIGURE 6.12   Idealization of steel decking for the Element Level Analysis.

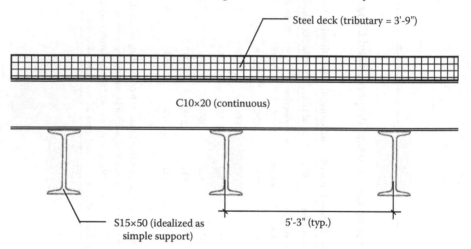

FIGURE 6.13   Idealization of steel channel (C) section for the Element Level Analysis.

Steel Channels

The following assumptions are applied to the steel channel for this level of analysis:

- Each channel (**C**) is continuously supported over several spans (i.e., constant spacing of 5'-3") with the supporting standard (**S**) sections idealized as simple supports (Figure 6.13); and
- The tributary area is bounded by the center to center spacing of the **C**-sections and **S**-section (3'-9" and 5'-3", respectively).

Steel Standards

The following assumptions are applied to the steel channel for this level of analysis:

- Each standard (S) section is idealized as a single-span beam with the supporting 36-in deep built-up members idealized as simple or fixed support depending on the type of connection to the built-up member. The simple connection is the critical one. (Figure 6.14); and

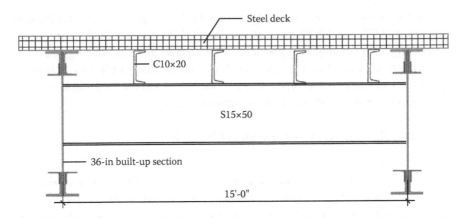

FIGURE 6.14   Idealization of steel standard (S) section for the Element Level Analysis.

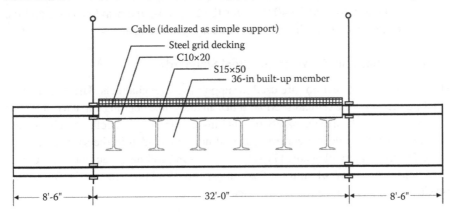

FIGURE 6.15   Idealization of 36-in deep built-up member.

- The tributary area is bounded by the center to center spacing of the S-sections and the built-up member (5'-3" and 15'-0", respectively).

Steel Built-Up Sections

The 36-in deep built-up section is represented by a beam with supporting cables idealized as simple supports (Figure 6.15). The tributary area of the vehicle traffic portion of the deck is bounded by the width of the bridge deck supported by the suspender cable and the center to center spacing of the built-up member (32'-0" and 15'-0", respectively). The tributary area of each overhang segment (or pedestrian portion) is bounded by the length of the overhang and the center to center spacing of built-up member (8'-6" and 15'-0", respectively).

MAXIMUM ALLOWABLE GROSS VEHICLE WEIGHT (GVW)

Maximum Allowable GVW on the Steel Decking

The commercially available open steel grid decking manufactured by the Interlocking Deck Systems International (IDSI), Inc.,[1] is comparable to the existing steel decking and is used in this case to determine the load capacity. The steel decking can carry a HS25 and HS20 truck at spacing of 5.01 ft and 5.34 ft, respectively. The existing steel decking is supported by Channel sections at spacing of 3.75 ft. It is therefore concluded that the steel decking will be able, at 0% loss in bending capacity, to carry any vehicle types shown in Tables 6.5 and 6.6, and will not control the determination of the allowable gross vehicle weight.

Maximum Allowable GVW on the Steel Channels

C10×20 sections are used to support the open grid steel decking. The A36 channel section has an allowable bending capacity of 31.3 k-ft and an allowable shearing capacity of 49 kips. Shear capacity, deflection limit, and connection capacity do not control and will not be included in the sample calculations. The maximum allowable GVWs at 0% loss in bending capacity (or $\eta = 0$) are: 34.38 tons, 31.98 tons, 33.96 tons, 43.26 tons, and 36.40 tons, for the 2-, 3-, 4-, and 5-axle trucks, and the 2-axle buses, respectively. Shear and deflection do not control.

Maximum Allowable GVW on the Steel Standards

S15×50 and S20×66 sections are used to support the steel channels. The A36 S15×50 standard section is the critical section. It has an allowable bending capacity of 95 k-ft and an allowable shearing capacity of 109 kips. Shear capacity, deflection limit, and connection capacity do not control and will not be included in the sample calculations. The maximum allowable GVWs at 0% loss in bending capacity (or $\eta = 0$) are: 42.50 tons, 39.53 tons, 41.98 tons, 53.49 tons, and 45.00 tons, for the 2-, 3-, 4-, and 5-axle trucks, and the 2-axle buses, respectively.

Maximum Allowable GVW on the Steel Built-Up Sections

Each A36 built-up member has an allowable bending capacity of 549.7 k-ft and an allowable shear capacity of 175 kips. Shear capacity, deflection limit, and connection capacity do not control and will not be included in the sample calculations. The maximum allowable GVWs at 0% loss in bending capacity (or $\eta = 0$) are: 20.17 tons, 18.77 tons, 19.93 tons, 36.68 tons, and 26.50 tons, for 2-, 3-, 4-, and 5-axle trucks, and the 2-axle buses, respectively.

CRITICAL MEMBER FOR DETERMINING GVW

The results from the Element Level Analysis indicate that the built-up member is the critical member for determining the load carrying capacity. In the following section, the results are generated for the built-up member.

SAMPLE CALCULATIONS FOR GVW LIMIT

The following illustrates how the maximum allowable GVW is determined for the critical member (i.e., built-up section):

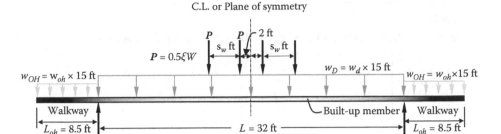

FIGURE 6.16   Loadings on 36-in built-up section.

Tributary Width, Length, and Area

Tributary width of the built-up member excluding the overhang = 15 ft
Tributary length of the built-up member excluding the overhang = 32 ft
Tributary area of the built-up member excluding the overhang = (15×32) ft^2
Tributary width of the built-up member overhang = 15 ft
Tributary length of the built-up member overhang = 8.5 ft
Tributary area of the built-up member overhang = (15×8.5) ft^2

Dead Loads

Open grid steel deck weight = 20 psf
Weight of other structural and non-structural components excluding overhang
   = 40 psf
Total dead weight excluding = w_d = (20 + 40) = 60 psf = 0.06 ksf
Dead weight on the overhang = w_{oh} ~ 50 psf = 0.05 ksf

Live Loads

Live load = Vehicle loading = Two trucks or buses placed side-by-side, separated by a distance of 4 ft (see Figure 6.16)

Live Load Distribution for the Front Axle
For the front axle, the load P in Figure 6.15.1 represents the resultant pressure under
the tire at one end of the front axle. Consequently, P is equal to 50% of the weight
attributed to the front axle, and can be represented by:

$$P = 0.5\xi W$$

where ξ = fraction of gross vehicle weight (GVW) attributed to the axle (Tables 6.5
and 6.6), and W = gross vehicle weight.

Live Load Distribution for the Rear Single and Tandem Axles
For the rear single axle, ξ = fraction of gross vehicle weight (GVW) attributed to the single rear axle (Truck Type 1 or Bus Type 1, 2, and 3 in Tables 6.5 and 6.6, respectively).
   For the rear tandem axle(s) for Truck Type 2 and 4 in Table 6.5, the centerline
of the tandem axles is placed over the built-up member. The percentage of the load

TABLE 6.8
Fraction of Gross Vehicle Weight, ξ Attributed to the Rear Axle(s)

Fraction of gross vehicle weight attributed to the rear axle(s)	Vehicle Type				
	Type-1	Type-2	Type-3	Type-4	Types 1, 2 & 3
ξ	0.8	0.75	0.67	0.44	0.67

distribution to the member is derived by considering a beam (**S** - section) in the longitudinal direction spanning between three built-up members with the centerline of the dual tandem axles placed on the built-up member in the middle. The built-up members are assumed to provide a simple support for the longitudinal beam.

For the three rear axles for Truck Type 3 in Table 6.5, the centerline of the middle rear axle is placed over the built-up member. The percentage of the load distribution to the member is derived by considering a beam (**S** - section) in the longitudinal direction spanning between three built-up members with the middle axle placed on the built-up member in the middle. The built-up members are assumed to provide a simple support for the longitudinal beam.

Table 6.8 presents the values of the fraction of GVW, ξ, attributed to the rear axle(s) for the trucks and buses in Tables 6.5 and 6.6, respectively.

For the rear axle(s), the load P can also be represented by:

$$P = 0.5\xi W$$

where ξ = fraction of gross vehicle weight (GVW) attributed to the rear axle(s) in Table 6.11, and W = gross vehicle weight.

Bending Moments

Moment due to dead load, M_D, (Figure 6.17):
Moment due to vehicle live load including a 30% impact load, M_{L+I}, (Figure 6.18):

Allowable GVW Calculation

Based on the allowable stress design (ASD), $M_a \geq M_D + M_{L+I}$ or $M_{L+I} \leq M_a - M_D$. When considering a loss in the allowable bending capacity (M_a) of magnitude η (where η = 19%, 38%, etc., Table 6.7), the moment relationship can be written as follows:

$$M_{L+I} \leq (1 - \eta)M_a - M_D \tag{5.1}$$

$$0.5 \times 1.3(L - s_w - 4)\xi W \leq (1 - \eta)M_a - \frac{w_D L^2}{8} + \frac{w_{OH} L_{oh}^2}{2} \tag{5.2}$$

$$W = \frac{(1-\eta) M_a - \frac{w_D L^2}{8} + \frac{w_{OH} L_{oh}^2}{2}}{0.5 \times 1.3 (L - s_w - 4)\xi} \tag{5.3}$$

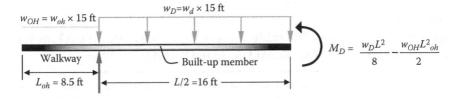

FIGURE 6.17   Moment due to dead loads.

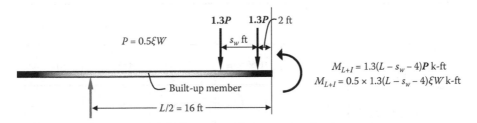

FIGURE 6.18   Moment due to truck or bus loads.

Considering that the built-up member has a 20% sectional loss [or loss in bending capacity of 38% (or $\eta = 0.38$) in Table 6.7] and is subjected to the 4-axle truck (Type 3 in Table 6.5), the maximum allowable gross vehicle weight (W) can be determined as follows:

$M_a = 549.7$ k-ft (Table 6.7 for 0% sectional loss)
$\eta = 0.38$ (38% loss in bending capacity)
$w_D = 0.9$ k/ft
$w_{OH} = 0.75$ k/ft
$L = 32$ ft
$L_{oh} = 8.5$ ft
$s_w = 6$ ft (Truck Type 3 in Table 6.5)
$\xi = 0.67$ (Truck Type 3 in Table 6.11)
$W = 26.38$ k = 13.19 tons

ALLOWABLE GVW OF TRUCKS AND BUSES

The allowable gross vehicle weights (GVWs) for the 2-, 3-, 4-, and 5-axle trucks, and the Type 1, 2, and 3 two-axle buses, are presented in Table 6.9 for different percentages of sectional losses varying from 10% to 40%, in 10% increments. The table is derived based on a deck self-weight of 20 psf.

ALLOWABLE GVW FOR DIFFERENT DECK WEIGHTS

In the event that replacement of the open grid deck will take place in the future, results are presented in Tables 6.10 to 6.14 for different deck weights (10 psf to 50 psf in 10 psf increments). The current deck weight is 20 psf.

TABLE 6.9

Element Level Analysis—Allowable Gross Vehicle Weight (GVW) in Tons for Different Percentages of Sectional Loss in the Built-Up Member

Vehicle Type	Allowable GVW (in tons) for different percentages of sectional loss*				
	0% Sectional Loss	10% Sectional Loss	20% Sectional Loss	30% Sectional Loss	40% Sectional Loss
2-axle truck-Type 1	20.17 tons	15.61 tons	11.04 tons	7.68 tons	4.80 tons
3-axle truck-Type 2	21.52 tons	16.65 tons	11.78 tons	8.19 tons	5.12 tons
4-axle truck-Type 3	24.09 tons	18.64 tons	13.19 tons	9.17 tons	5.73 tons
5-axle truck-Type 4	36.68 tons	28.38 tons	20.08 tons	13.97 tons	8.72 tons
2-axle bus - Types 1, 2, & 3	26.50 tons	20.50 tons	14.51 tons	10.09 tons	6.30 tons

*   In case a % sectional loss falls between two values (e.g. 14% sectional loss), a linear interpolation between the % sectional loss that is lower and the one that is higher than the one in question (e.g. 10% and 20% sectional loss) should yield adequate results.

SUMMARY AND CONCLUSIONS

The primary objective of the structural evaluation of the John A. Roebling Bridge is to determine the maximum allowable gross vehicle (truck or bus) weight (GVW) that can be carried by the bridge deck structural elements: steel grid decking, channels, standard sections, and/or built-up members. The John A Roebling Bridge carries KY 17 over the Ohio River between Covington, KY, and Cincinnati, OH. A detailed evaluation of the load carrying capacity of the cables and truss elements was completed in 2003.[4]

An "Element Level Analysis" is carried out to determine the maximum allowable GVW for different truck and bus types. The bridge deck structural elements are analyzed independent of each other. Each element is assigned a specific tributary area, and the element support conditions are idealized as appropriate (i.e., simple, fixed, etc).

Four truck types and three bus types are considered in the analysis. In 2007, the posted weight limits on the bridge were 17 tons for two-axle trucks and 22 tons for three-, four-, and five-axle trucks.

The built-up 36 inch deep member turned out to be the critical member. The maximum allowable GVWs for trucks and buses are presented in Table 6.15, which assumes a deck self-weight of 20 psf.

TABLE 6.10

Element Level Analysis—Allowable Gross Vehicle Weight (GVW) in Tons for Different Percentages in Sectional Loss in the Built-Up Member When the Deck Weight Equals 10 psf

	Deck Weight = 10 psf				
	Allowable GVW (in tons) for different percentages of sectional loss*				
Vehicle Type	**0% Sectional Loss**	**10% Sectional Loss**	**20% Sectional Loss**	**30% Sectional Loss**	**40% Sectional Loss**
2-axle truck-Type 1	21.01 tons	16.45 tons	11.88 tons	8.52 tons	5.64 tons
3-axle truck-Type 2	22.41 tons	17.55 tons	12.68 tons	9.09 tons	6.01 tons
4-axle truck-Type 3	25.09 tons	19.64 tons	14.19 tons	10.17 tons	6.73 tons
5-axle truck-Type 4	38.21 tons	29.91 tons	21.61 tons	15.49 tons	10.25 tons
2-axle bus - Types 1, 2, & 3	27.60 tons	21.60 tons	15.61 tons	11.19 tons	7.40 tons

* In case a % sectional loss falls between two values (e.g. 14% sectional loss), a linear interpolation between the % sectional loss that is lower and the one that is higher than the one in question (e.g. 10% and 20% sectional loss) should yield adequate results.

In the event that the existing open grid deck were to be replaced, results in anticipation of such replacement are presented for different deck weights (10 psf to 50 psf, in 10 psf increments) in section 5. The current open grid deck weight is approximately 20 psf.

REFERENCES

1. Interlocking Deck Systems International (IDSI), Inc. (2004): http://www.idsi.org/
2. AASHTO, "Standard Specifications for Highway Bridge. 17th Ed. American Association of State Highway and Transportation Officials. Washington, D.C., 2002.
3. AISC, "Steel Construction Manual," 13th Edition, American Institute of Steel Construction Inc., 2005.
4. Ren, W.X., Harik, I.E., Blandford, G.E., Lenett, M., and Baseheart, T.M., "Structural Evaluation of The Historic John A. Roebling Suspension Bridge," Research Report (KTC03-10/MSC97-1F), Kentucky Transportation Center, University of Kentucky, April 2003.

TABLE 6.11

Element Level Analysis—Allowable Gross Vehicle Weight (GVW) in Tons for Different Percentages in Sectional Loss in the Built-Up Member When the Deck Weight Equals 20 psf

	Deck Weight = 20 psf				
Vehicle Type	Allowable GVW (in tons) for different percentages of sectional loss*				
	0% Sectional Loss	**10%** Sectional Loss	**20%** Sectional Loss	**30%** Sectional Loss	**40%** Sectional Loss
2-axle truck-Type 1	20.17 tons	15.61 tons	11.04 tons	7.68 tons	4.80 tons
3-axle truck-Type 2	21.52 tons	16.65 tons	11.78 tons	8.19 tons	5.12 tons
4-axle truck-Type 3	24.09 tons	18.64 tons	13.19 tons	9.17 tons	5.73 tons
5-axle truck-Type 4	36.68 tons	28.38 tons	20.08 tons	13.97 tons	8.72 tons
2-axle bus - Types 1, 2, & 3	26.50 tons	20.50 tons	14.51 tons	10.09 tons	6.30 tons

* In case a % sectional loss falls between two values (e.g. 14% sectional loss), a linear interpolation between the % sectional loss that is lower and the one that is higher than the one in question (e.g. 10% and 20% sectional loss) should yield adequate results.

TABLE 6.12

Element Level Analysis—Allowable Gross Vehicle Weight (GVW) in Tons for Different Percentages in Sectional Loss in the Built-Up Member When the Deck Weight Equals 30 psf

Vehicle Type	Deck Weight = 30 psf				
	Allowable GVW (in tons) for different percentages of sectional loss*				
	0% Sectional Loss	10% Sectional Loss	20% Sectional Loss	30% Sectional Loss	40% Sectional Loss
2-axle truck-Type 1	19.34 tons	14.77 tons	10.21 tons	6.84 tons	3.96 tons
3-axle truck-Type 2	20.62 tons	15.76 tons	10.89 tons	7.30 tons	4.22 tons
4-axle truck-Type 3	23.09 tons	17.64 tons	12.19 tons	8.17 tons	4.73 tons
5-axle truck-Type 4	35.16 tons	26.86 tons	18.56 tons	12.44 tons	7.20 tons
2-axle bus - Types 1, 2, & 3	25.40 tons	19.40 tons	13.40 tons	8.99 tons	5.20 tons

* In case a % sectional loss falls between two values (e.g. 14% sectional loss), a linear interpolation between the % sectional loss that is lower and the one that is higher than the one in question (e.g. 10% and 20% sectional loss) should yield adequate results.

TABLE 6.13

Element Level Analysis—Allowable Gross Vehicle Weight (GVW) in Tons for Different Percentages in Sectional Loss in the Built-Up Member When the Deck Weight Equals 40 psf

	Deck Weight = 40 psf				
Vehicle Type	Allowable GVW (in tons) for different percentages of sectional loss*				
	0% Sectional Loss	10% Sectional Loss	20% Sectional Loss	30% Sectional Loss	40% Sectional Loss
2-axle truck-Type 1	18.50 tons	13.93 tons	9.37 tons	6.00 tons	3.12 tons
3-axle truck-Type 2	19.73 tons	14.86 tons	9.99 tons	6.40 tons	3.33 tons
4-axle truck-Type 3	22.09 tons	16.63 tons	11.18 tons	7.17 tons	3.73 tons
5-axle truck -Type 4	33.63 tons	25.33 tons	17.03 tons	10.91 tons	5.67 tons
2-axle bus - Types 1, 2, & 3	24.29 tons	18.30 tons	12.30 tons	7.88 tons	4.10 tons

* In case a % sectional loss falls between two values (e.g. 14% sectional loss), a linear interpolation between the % sectional loss that is lower and the one that is higher than the one in question (e.g. 10% and 20% sectional loss) should yield adequate results.

TABLE 6.14
Element Level Analysis—Allowable Gross Vehicle Weight (GVW) in Tons for Different Percentages in Sectional Loss in the Built-Up Member When the Deck Weight Equals 50 psf

Vehicle Type	Deck Weight = 50 psf				
	Allowable GVW (in tons) for different percentages of sectional loss*				
	0% Sectional Loss	10% Sectional Loss	20% Sectional Loss	30% Sectional Loss	40% Sectional Loss
2-axle truck-Type 1	17.66 tons	13.09 tons	8.53 tons	5.16 tons	2.28 tons
3-axle truck-Type 2	18.83 tons	13.97 tons	9.10 tons	5.51 tons	2.43 tons
4-axle truck-Type 3	21.08 tons	15.63 tons	10.18 tons	6.17 tons	2.72 tons
5-axle truck-Type 4	32.10 tons	23.80 tons	15.50 tons	9.39 tons	4.15 tons
2-axle bus - Types 1, 2, & 3	23.19 tons	17.20 tons	11.20 tons	6.78 tons	3.00 tons

* In case a % sectional loss falls between two values (e.g. 14% sectional loss), a linear interpolation between the % sectional loss that is lower and the one that is higher than the one in question (e.g. 10% and 20% sectional loss) should yield adequate results.

TABLE 6.15

Allowable Gross Vehicle Weight (GVW) in Tons for Different Percentages of Sectional Loss in the Built-Up Member (Assuming a Deck Self-Weight of 20 psf)

Vehicle Type	Allowable GVW (in tons) for different percentages of sectional loss*				
	0% Sectional Loss	10% Sectional Loss	20% Sectional Loss	30% Sectional Loss	40% Sectional Loss
2-axle truck-Type 1	20.17 tons	15.61 tons	11.04 tons	7.68 tons	4.80 tons
3-axle truck-Type 2	21.52 tons	16.65 tons	11.78 tons	8.19 tons	5.12 tons
4-axle truck-Type 3	24.09 tons	18.64 tons	13.19 tons	9.17 tons	5.73 tons
5-axle truck-Type 4	36.68 tons	28.38 tons	20.08 tons	13.97 tons	8.72 tons
2-axle bus - Types 1, 2, & 3	26.50 tons	20.50 tons	14.51 tons	10.09 tons	6.30 tons

* In case a % sectional loss falls between two values (e.g. 14% sectional loss), a linear interpolation between the % sectional loss that is lower and the one that is higher (e.g. 10% and 20% sectional loss) should yield adequate results.

7 Extant Lenticular Iron Truss Bridges from the Berlin Iron Bridge Company

Alan J. Lutenegger

CONTENTS

ABSTRACT

Of the estimated five to six hundred lenticular iron truss bridges built by the Berlin Iron Bridge Company over a twenty year period from about 1880 to 1900 only about 50 bridges remain. Extant bridges are located in Hew Hampshire, Vermont, Massachusetts, Connecticut, Rhode Island, New York, New Jersey, Pennsylvania and Texas. Over the past 5 years, the author has visited all but two of the extant bridges and documented their geometry and condition. Some of the bridges are still in service, some are closed to traffic, some have been adapted for use as pedestrian bridges and some are in storage awaiting a final decision on their fate. These bridges are an example of a special type of catalog bridge available in the mid to late 19th Century. The bridges represent a unique design during a time when many bridge companies were competing for contracts in a highly competitive market. These bridges were the only ones to use the lenticular design by a prominent bridge builder of the era. A background on the lenticular design is given and a statistical summary of the geometry of the remaining bridges is given. The data collected provide insight into the design and applicability of the bridges that were selected at specific locations.

INTRODUCTION

The author visited all but four of the known remaining lenticular iron truss bridges built by the Berlin Iron Bridge Company (BIBCO) to gather information on the specific geometry and construction of the various bridge components as an initial phase in evaluating the engineering approach used for these bridges. The bridges range in age from 1878 to 1896 and therefore essentially represent the full time period over which the bridges were built. Pony truss bridges and through truss bridges including suspended decks, mid deck and under deck designs are represented in the remaining bridges. A total of 57 locations were visited and a total of 69 structures were examined since there are some locations with more than one structure. All but seven of the known structures have been examined at this time. The author also visited the BIBCO lenticular bridges in Texas, which may have been the only lenticular bridges built west of the Mississippi River. Even though the number of extant bridges represents only about 15% of the estimated total number of bridges produced, there appears to be a sufficiently good representation of nearly every style of bridge produced by the company.

Darnell (1979) presented a partial listing of surviving lenticular truss bridges and at that time included 48 locations (total of 57 individual spans). The list did not include any bridges in New Hampshire, Rhode Island, Pennsylvania or the bridges in Texas. Unfortunately, a number of the bridges in Darnell's 1979 list have been removed and demolished but a number of other bridges have been rediscovered and are included in the list in this paper. A description of the extant lenticular bridges in Massachusetts has recently been given by Lutenegger and Cerato (2005). At this writing bridges are known to exist in eight states, primarily in New England as well as Texas. Recent renewed interest in these bridges has produced a new review of their structural analysis (Boothby 2004). In this paper, an examination is made of the configuration of the remaining bridges including an evaluation of the overall shape as well as a look at some individual components. No attempt has been made to perform a detailed structural analysis. Rather, an initial analysis of the dimensions of the bridges has been performed to allow some insight into the process used by the designers to select specific geometries and components at various locations.

LENTICULAR TRUSS BRIDGES

During the latter part of the 19th Century and the beginning of the 20th Century, the Berlin Iron Bridge Co. of East Berlin, Connecticut, manufactured and erected almost 400 lenticular truss bridges in the United States (Darnell 1979). These bridges are sometimes referred to as "pumpkin-seed bridges", "fish-belly bridges", "cats-eyes bridges", "elliptical truss bridges", "double bowstring", or "parabolic truss bridges" because of their unique lens shape. Like many other iron truss bridges of the day,

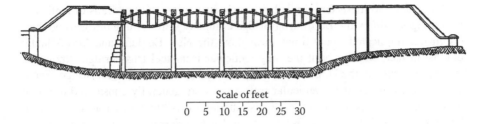

Scale of feet

0   5   10   15   20   25   30

FIGURE 7.1   Stockton & Darlington Railway Bridge 1823.

these bridges were, in effect, "mass produced" as the components were built in a factory, sent to the site, and then assembled. Many of the components were used repeatedly for different spans or applications.

According to James (1981) lenticular shaped bridges had previously been used in Europe as early as 1822. Wooden bridges with a lenticular shape had been used in Germany by Laves. It appears that one of the first uses of this type of design in iron was George Stephenson's railway bridge designed in 1822 and built between 1823 and 1824 to carry the Stockton & Darlington Railway over the river Gaunless in West Auckland, UK. As shown in Figure 7.1, the bridge consisted of four spans of 12 ½ ft. (originally three spans with a fourth span added in 1825) with top and bottom chords of wrought iron and the vertical members of cast iron. The members were built by Burrell & Company of Newcastle. The bridge was opened on September 27th, 1825 and was in use until about 1856. The bridge stood intact but unused until 1901, when it was dismantled and moved to storage. In 1928 the bridge was re-erected at the York Railway Museum and is currently on display at the UK National Railway Museum. The author has recently been allowed to examine and document this bridge which is still in remarkably good shape.

One of the most notable bridges of this style was I.K. Brunel's 1855 twin span lenticular Royal Albert Railway Bridge across the Tamar in Saltash UK. This bridge used tubular upper chords with each span having a span of 445 ft. (center to center of the piers). In 1860, the Mainz Bridge was built over the Rhine River in Germany and consisted of at least two large spans and two shorter spans. This bridge shows remarkable similarities in form to the lenticular truss bridges built twenty-five years later by the Berlin Iron Bridge Company and discussed later in this paper.

In the U.S., the great bridge engineer Gustav Lindenthal built a lenticular shaped twin span bridge across the Monongahela River at Smithfield Street in Pittsburgh, PA, in 1883. Lindenthal (1883) referred to this shape as a "Pauli Truss" after the famous German bridge engineer Friedrich August von Pauli (1802–1883). Each span of the bridge originally was constructed using two trusses; a third truss was added to carry streetcar traffic in 1891. This bridge replaced an earlier one designed and built by John Roebling, but did not really receive the attention that perhaps Lindenthal had been hoping for. This may have been in part related to the fact that another bridge which opened in 1883, the Brooklyn Bridge, may have had considerably more importance at the time. However, the structures of Brunel and Lindenthal were unique, single event, monumental bridges, never to be duplicated in any close form by any other engineer at any other location.

By contrast to these few single large scale structures, the hundreds of lenticular truss bridges built by the Berlin Iron Bridge Company were catalog bridges and their design was duplicated many times throughout the New England and Mid-Atlantic regions. In fact, BIBCO built the only lenticular iron/steel truss bridges known to have been erected in the U.S. aside from Lindenthal's Smithfield Street Bridge. These bridges were only used for vehicular traffic and were generally considered too light to be used for railroad and trolley loads, although it is known that at least one, in Portland, Maine, did also serve as a trolley bridge. Considering that the most common traffic of the era (1880–1900) consisted of horse-drawn carts or wagons, it is amazing that any of the bridges survived through the automobile age to the present day. Most of the bridges that were lost over the years were not because of failures from overloading; most were swept away during severe floods.

Darnell (1979) described a number of lenticular bridges and gives a detailed account of the history of the Berlin Iron Bridge Co. In addition to the uniquely shaped lenticular truss bridges, BIBCO also built conventional steel truss bridges and even built a few pedestrian suspension bridges, the most notable of which were erected in Keesville, N.Y. and Milford, N.H. In addition to bridges, BIBCO had a thriving business, building roof trusses, water towers, and complete steel frames for buildings. The company was very persistent in its advertising and routinely placed ads in a number of important and influential trade magazines and journals of the day, including the Transactions of the ASCE.

THE PATENTS OF WILLIAM O. DOUGLAS

A patent, No. 202,526, was issued by the U.S. Patent Office on April 16, 1878 to William O. Douglas, of Binghamton, N.Y., for a truss bridge, described in the patent as "A combination of two or more elliptical trusses connected as herein described with the floor and joints and necessary flooring to form a through deck or swing bridge". Douglas's patent drawings of 1878 included only three basic configurations of bridges, namely, suspended deck, under deck and mid deck bridges, as shown in Figure 7.2. Two of Douglas's patent drawings showing a suspended deck design (i.e., deck tangent to the lower chord) are shown in Figure 7.2.

Earlier patents for bridges with similar features to those of Douglas's (especially the parabolic or elliptical shape) had previously been issued by the U.S. Patent Office; on March 27, 1849 to James Barnes (No, 6,230); on September 2, 1851, to Edwin Stanley, N.Y. (No. 8,337); on August 21, 1855 to Horace L. Hervey and Robert E. Osborn (No. 13,461); on March 28, 1871 to Ferdinand Dieckman (No. 113,030); on October 22, 1872 to G.E. Harding (No. 132,398); and on June 11, 1873 to James B. Eads (No. 142, 381). Even after Douglas received his patent and the Berlin Iron Bridge Company had been in full production for many years, the U.S. Patent Office continued to grant patents for bridge designs with noted similarities to previous structures, for example on December 9, 1884 to Charles Strobel (No. 309,171); and on August 18, 1896 to John Semmes (No. 566,233).

Douglas had published his suggestion for his bridge design in an 1877 printing of the *Scientific American Supplement*, showing an illustration of his proposed design, Figure 7.3. Douglas (1877) referred to his design as "An Elliptical Truss Bridge"

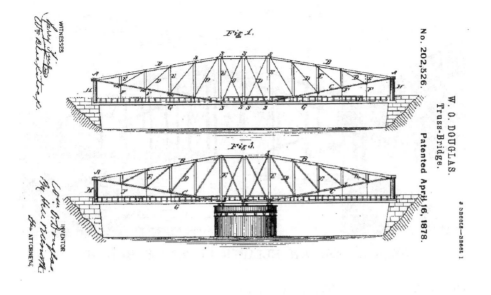

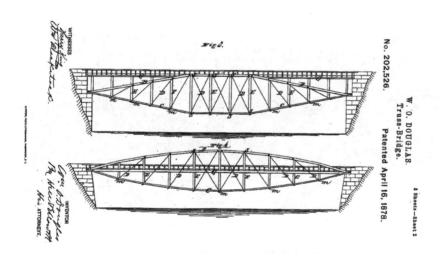

FIGURE 7.2   Drawings from William O. Douglas's 1878 patent.

noting that "In a bridge as above illustrated in Figure 7.2, we have the arch and suspension principles united, forming an elliptical truss. The thrust of the arch equipoises and is equipoised by the pull of the cable". He further noted that "The roadway is suspended to the two chords so that the arch carries one half of the load and the cable the other, under which circumstances the thrust and the pull at the top of end posts will be equal. The end posts have only to support the dead load of the bridge". Douglas's public disclosure of his proposed design (July 14, 1877) predates

DESIGN FOR AN ELLIPTICAL TRUSS BRIDGE.

FIGURE 7.3   Design for elliptical truss bridge.
Source: Douglas (1877).

DESIGN FOR A COMPOUND TRUSS IRON BRIDGE.

FIGURE 7.4   Suggestion for a compound truss bridge.
Source: Douglas (1877).

his patent application (March 28, 1878) by nearly eight months. This is the only known publication by Douglas or any other engineer associated with the Berlin Iron Bridge Company or its predecessor, the Corrugated Metal Company, related to the lenticular design during this era.

Douglas (1877) also suggested a "Compound" truss bridge as shown in Figure 7.4. At the end of his brief note, Douglas made a general appeal for comments "Criticisms of the general principles of this elliptical truss, positive and comparative, are respectfully invited from engineers over their signature". There are no published responses to Douglas's request in subsequent issues of *Scientific American Supplement*. At least one surviving bridge builder's plate (Depot Rd. No. Chichester, N.H.) indicates "Douglas & Jarvis Patents, April 16, 1878, April 5, 1885". There are no known patents related to the lenticular shape attributed to Jarvis.

BRIDGE PRODUCTION BY THE BERLIN IRON BRIDGE COMPANY

A number of the bridges had been built by the predecessor of the Berlin Iron Bridge Company, the Corrugated Metal Company, out of their small manufacturing plant located in Berlin, Connecticut. It appears that one of the first lenticular bridges by the Corrugated Metal Company was a four-panel Pony Truss bridge apparently built in 1879 and erected at Waterbury, Connecticut, spanning the Naugatuck River. This bridge is still standing and is in use as a vehicular bridge. The smallest of these bridges consisted of 3 panels, with a span of only about 40 ft. Bridges were built principally throughout the northeast, with surviving examples that still may be seen in Massachusetts, Vermont, New Hampshire, Rhode Island, Connecticut, New York, New Jersey and Pennsylvania. It is also known that several bridges were built in Ohio, but it seems that none survive. Interestingly enough, there are at least six extant lenticular truss bridges in Texas, the only ones known to have been sold and built west of the Mississippi River, and thought to have been the work of an extremely enthusiastic free lance salesman in Texas. There is no indication that any bridges were actually built with the simple configuration of Douglas's suggestion of 1877. A few short span 3 panel pony truss bridges do show some similarity to his drawing such as at Sharon Station, Ct., Hadley, N.Y. and Smithfield, R.I. The span lengths of these remaining three panel pony truss bridges are 40 ft., 44.5 ft., and 48 ft., respectively.

The name of the Corrugated Metal Company was changed to the Berlin Iron Bridge Company sometime around 1883 according to Darnell (1979) and, according to company literature, they provided almost 90% of the iron bridges roadway bridges throughout New England from 1880 to 1890. Designs for the bridges included both Pony Truss and Through Truss configurations. A second U.S. patent (No. 315,259) was granted to Douglas for improvements on his design on April 7, 1885. The primary improvements that Douglas incorporated into this patent were the use of floor line tension chords and strut braces. The strut brace was noted by Douglas to improve the bridges behavior under wind loading. The floor line tension chord was often simply a wrought iron rod running the length of the truss and connected to the end posts on either end. A turnbuckle was used to adjust the tension. Douglas's 1885 patent drawings show only his strut brace and floor line chord applied to a suspended deck bridge.

Figure 7.5 shows the typical bridge configurations used by the BIBCO at different locations. Most of the extant lenticular truss bridges are of the suspended deck style. There is only one mid-deck bridge and only two under deck spans known to exist. Figure 7.6 shows the various combinations of bridges used at different locations.

Douglas died around 1890, but the Berlin Iron Bridge Co. continued to be very productive under the leadership of several men. Early ads run by the company indicate the

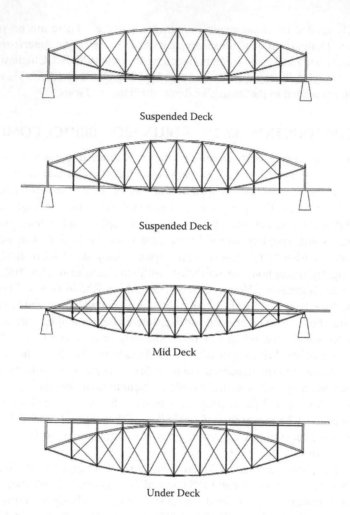

Suspended Deck

Suspended Deck

Mid Deck

Under Deck

FIGURE 7.5   Bridge configurations constructed by the Berlin Iron Bridge Company.

following principals: Charles M. Jarvis, President and Chief Engineer; Mace Moulton, Consulting Engineer; Burr K. Field, Vice President; George H. Sage, Secretary; and F. L. Wilcox, Treasurer.

The Berlin Iron Bridge Company, under the leadership of Charles M. Jarvis, acquired the rights to Douglas's patent, which accounts for the exclusive promotion of this style of bridge by the company. Additionally, the company apparently had excellent salesmen or agents who were most likely paid on commission, many of whom may have had a special affinity to this style of bridge, especially since the company designed and built other more conventional truss bridges. No other lenticular bridges built by any other bridge manufacturer of the era are known to have been designed, built or even advertised. This style of bridge was unique to the Berlin Iron Bridge Company.

A number of textbooks available during the period from about 1870 to 1885 presented descriptions of this style of bridge, referred to at that time as "Double

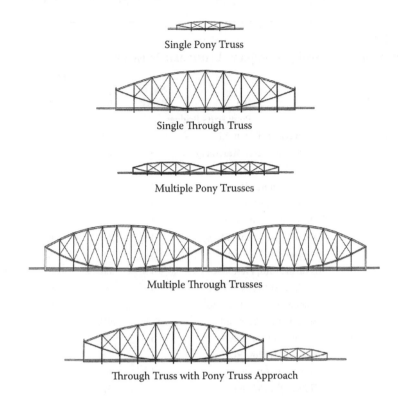

Single Pony Truss

Single Through Truss

Multiple Pony Trusses

Multiple Through Trusses

Through Truss with Pony Truss Approach

FIGURE 7.6   Different bridge combinations used at various locations.

Bowstring", "Parabolic", "Bowstring Suspended" or "Lenticular" bridges (e.g., Campin 1871; Matheson 1873; Shreve 1873; Rankine 1874; Ritter 1879; DuBois 1883). Even after the end of the 19th century and into the 1920s, the bridges were still discussed in some texts (e.g., Merriman and Jacoby 1906; 1920). In Chapter X of his book, Shreve (1873) provided an example of the design of a lenticular truss, described by Shreve as "The form of this peculiar truss, known also as the Pauli System..." noting that "It is not capable of supporting any greater weight than a Bow String Truss of equal depth and length, and practically possesses many disadvantages". DuBois (1883) shows a diagram of a lenticular truss with Warren style bracing noting that "the bracing may be of any sort" and also that "In Germany, this shape is known as the *Pauli* truss".

SUMMARY OF SURVIVING BRIDGES

Throughout New England and other parts of the U.S., there are a number of surviving lenticular bridges built by the Berlin Iron Bridge Co. A few have been restored; many are closed to traffic; some have been adapted for other uses, such as pedestrian bridges, while others have already been scheduled for replacement. A summary of the remaining lenticular truss bridges known to the author is given in Table 7.1. By far, the most common style of bridges produced in either the pony truss or through truss configuration were simple suspended deck bridges; these are represented most by the

TABLE 7.1
Extant Berlin Iron Bridge Company Lenticular Truss Bridges

No.	State	Town	No. of Panels	Span (ft)	Mid-Span Height (ft.)	Aspect Ratio, L/H
		Pony Truss Bridges				
1	Connecticut	Main Street, Talcottville	4	61	8	7.6
2		Washington Ave., Waterbury	5	67.5	8	8.4
3		Sheffield Street, Waterbury	4	53	8	6.6
4		Minortown Rd., Woodbury	4	63	8	7.9
5		W. Main Street, Stamford	5	60.5 (2)	10	6.0
6		Oliver Street, Stamford	4	53	8	6.6
7		Melrose Ave., E. Windsor	4	63	8	7.9
8		Oregon Rd., Meriden	5	78	8	9.7
9		Hallville Mill, Preston	4	60	6	10.0
10		Ashland Mill, Jewett City	4	65	6	10.8
11		Sharon Station Rd., Sharon	3	40	6	6.7
12		Sebethe Dr., Cromwell	N/A	N/A	N/A	N/A
13	Massachusetts	Golden Hill Rd., Lee	5	80	8	10.0
14		Pumpkin Hollow Rd.	4	58	8	7.2
15		Fort River, Amherst	4	60	8	7.5
16		Gilbert, Rd., Warren	5	72	8	9.0
17		Blackstone Bikeway	6	74	8	9.2
18		North Canal, Lawrence	5	83	8	10.4
19	New Hampshire	Delage Rd., Franconia	4	56	6	9.3
20		Dow Ave., Franconia	5	69	8	8.6
21	New Jersey	Cleveland Bridge, Mahwah	5	83	8	10.4
22	New York	Water St., Homer (east)	4	45	6	7.5
23		Water St., Homer (west)	4	52.5	6	8.7
24		Wall St., Homer	4	52.5	6	6.6
25		Pine St., Homer	4	58	8	7.2
26		Rhule Rd, Malta	5	71	6	11.8
27		Corinth Rd., Hadley	3	44.5	6	8.9
28		Silk St., Newark Valley	5	69	8	8.6
29		Avoca (Wallace)	4	51	N/A	—
30		Niobi Junction	3	N/A	N/A	—
31		Long Lane Road, Steuben Co.	4	N/A	N/A	—
32	Pennsylvania	S. High St., Duncannon	5	81	8	10.1
33	Rhode Island	Stillwater Rd., Smithfield	3	48	6	8.0
34		Interlaken Mills, Arkwright	6	91	11.33	8.0
35	Texas	Augusta St., San Antonio	7	98	22.2	4.4
36		Crockett St., San Antonio	6	84	13.7	6.1
37		S. Presa St., San Antonio	7	98	13.7	7.2
38		Soda Spring Rd.	5	75	8	9.4

TABLE 7.1 (continued)
Extant Berlin Iron Bridge Company Lenticular Truss Bridges

No.	State	Town	No. of Panels	Span (ft)	Mid-Span Height (ft.)	Aspect Ratio, L/H
		Pony Truss Bridges (continued)				
39		Salado	6	87	8	10.9
40		Kelley Crossing, Caldwell Co	6	89	12	7.4
41	Vermont	Randolph Rd., Bethel	4	58	8	8.2
42		Highgate Falls	5	68.75	8	8.6
		Through Truss Bridge				
43	Connecticut	Brunswick Ave, Plainfield	8	124	19	6.5
44		Main St., Moosup	7	105	19	5.5
45		Lover's Leap, New Milford	10	172.5	30	5.8
46		Boardman's Rd., New Milford	13	188	30	6.3
47	Massachusetts	Galvin Rd., N. Adams	7	103	18	5.7
48		Bardwell's Ferry Rd., Shelburne	13	198	30	6.6
49		Aiken St., Lowell	11	153 (5)	32.8	4.6
50	New Hampshire	Depot Rd., N. Chichester	6	95	19.66	4.8
51	New Jersey	Neshanic Station	9	140.5 (2)	23.0	6.1
52	New York	Kelsey St., Candor	5	69	16.33	4.2
53		Ouaquaga, Colesville	8	170 (2)	30	5.7
54		Washington St., Binghampton	10	170 (3)	30	5.7
55		Taylor Town Line Rd.	6	82.5	17.33	4.8
56		Delhi	8	114.5	20.5	5.6
57		Walton Bridge, Keene	N/A	N/A	N/A	—
58	Texas	Yancey Rd., Frio Co. (now Santa Cruz Ranch)	7	109	19.83	5.5
59		Brackenridge Park, San Antonio	7	92.75	23	4.0
60	Pennsylvania	Pierceville, Nicholson Township, Wyoming Co.	8	113.83	16	7.6
61	Vermont	Highgate Falls	13	214.5	30	7.1
		Through Truss—Warren Configuration				
62	New York	Raymondville	9	288	36.5	7.9
63	Pennsylvania	Jersey Shore, Lycoming Co.	14	287	39.66	7.2
64		Waterville, Lebanon Co	12	220.75	33	6.7
		Half-Deck				
65	New York	Corinth Rd., Hadley	9	136	18	7.6
		Under Deck				
66	New Hampshire	Livermore Falls, Compton	10	138	30.2	4.6
67		Livermore Falls, Compton	7	101.5	14	7.2

Note: Number in ( ) indicated number of spans.

extant bridges. Of the known bridges at this writing, there are a total of 43 pony truss bridges, a total of 27 through truss bridges, three through truss bridges with Warren bracing, one mid-deck bridge and two under deck bridges. While a complete analysis of the variations of all the different bridge components is beyond the scope of this paper, a number of initial observations may be made for both pony truss and through truss bridges. The Texas bridges are not discussed in this paper as they are considered by the author as a unique set of bridges, especially those in San Antonio.

ANALYSIS OF SURVIVING BRIDGES

Aspect Ratio (L/H)

Included in Table 7.1 is the measured span length of each bridge, taken as the distance between center of pins on the end posts, and the mid-span height of each bridge, taken as the distance between pins on the center vertical hanger. The aspect ratio of the bridges, taken as the span length/mid span height (L/H), is also given. The aspect ratio of the pony truss bridges ranges from about 6 to 11 and is dictated by the fact that the maximum height at mid-span of any of the bridges is only 10 ft. This gives the pony truss bridges the appearance of a long and slender span. Nearly all pony truss bridges have either a 6 ft. or an 8 ft. vertical post at mid-span, which appear to be standard components used by the Company. The 6 ft. posts were often used on bridges with a mid-span post of 8 ft. The 6 ft. and 8 ft. vertical web posts occur in both a tapered configuration and a parallel configuration as will be discussed. Other pony truss bridges appear to represent special designs; these include the twin spans of West Main Street in Stamford, Ct., the Interlaken Mills Bridge in Arkwright, R.I and the Kelley Crossing Bridge in Texas. The latter two bridges are nearly identical bridges of the "half through" design; that is, with the deck supported above the tangent horizontal line to the lower chord. These are the only known examples of this style bridge to survive. All four of these spans have a higher mid-span height as compared to all of the other pony truss bridges.

Figure 7.7 shows a plot of the aspect ratio as a function of the span length for all pony truss bridges, including those located in Texas. Since most of the bridges throughout the northeast are what can be considered as a standard design and have either a mid-span web post of either 6 ft. or 8 ft. in height, it can be seen that the aspect ratio increases linearly with span length. Two of the Texas brides (Soda Spring Rd. and Salado) were essentially designed like all of the bridges in the northeast and fit into the lower linear trend line. The other four Texas bridges, all of which are located in San Antonio, show individual unique configurations. They are among the longest of the pony truss bridges constructed and they have relatively large mid-span heights. There is considerable overlap in the two linear trend lines which results from designers using both 6 ft. and 8 ft. mid-span web posts on bridges of the same span length. The transition is essentially from 3 to 4 to 5 panels for all the standard designs. Outside of Texas, there are only 2 known pony truss bridges with more than 5 panels.

The aspect ratio of the through truss bridges ranges from 4 to about 8 for span lengths ranging from 69 ft. to 215 ft. This gives many of the through truss bridges a more rounded "elliptical" shape when viewed from the side. Figure 7.8 shows the

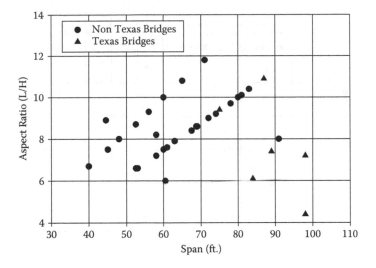

FIGURE 7.7   Relationship between span length and aspect ratio—pony trusses.

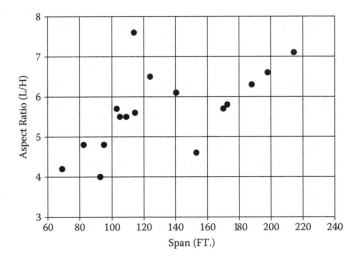

FIGURE 7.8   Relationship between span length and aspect ratio—through trusses.

variation of aspect ratio with span length for all of the through truss bridges (not including the three Warren configurations). There appears to be considerably more scatter in these dimensions than in the pony truss, especially in the shorter span lengths, from about 69 ft. to 120 ft. It is possible on the longer span bridges, i.e., after about 150 ft. that designers started to use a fixed maximum mid-span height of 30 ft.

NUMBER OF PANELS

Figure 7.9 shows the variation in the number of panels used in the construction as a function of the span length for both pony truss and through truss bridges (not including Texas bridges or Warren bridges). The overlap is more apparent in the pony truss

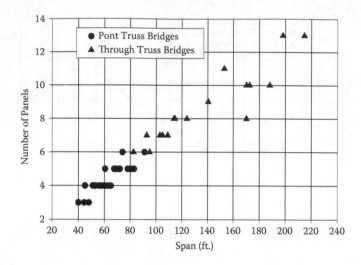

FIGURE 7.9 Variation in number of panels as a function of span length.

designs. As the required span length increased designers simply extended the existing panels in some cases while in other cases they added a panel.

VERTICAL WEB POSTS

Vertical web posts used on the pony truss bridges to connect the upper and lower chords are the simplest of all of the built-up members and were fabricated from four angle sections with riveted flat bar diagonals. Web posts were either tapered or were constructed with parallel sides as shown in Figure 7.10. Tapered web posts were connected to the pins at the upper chords on the inside of the chord, while parallel web posts were connected to the upper chords on the outside of the chord. It appears that these components for suspended deck pony truss bridges were mass produced to a standard configuration at the East Berlin plant and then the geometry of other components were adjusted to fit the required span length needed at individual sites.

For example, not including the Texas bridges which appear to be unique unto themselves, essentially all pony truss bridges used vertical posts with a maximum fixed length of 8 ft. Regardless to the number of panels used to assemble the full span, vertical posts of 8 ft. and 6 ft. were almost exclusively used to produce 3, 4, and 5 panel bridges with span lengths ranging from 40 ft. to 83 ft., as indicated in Table 7.1. The ends of the top chords could have easily been adjusted to give the appropriate length and connecting angle so that riveted plates would fit properly between chord members. This means that the vertical posts could be assembled and placed in inventory until the other components were fabricated. Since the bottom chord of all bridges consisted on wrought iron flat eye bars, their lengths could also have been easily adjusted to fit the specific location. Table 7.2 gives a summary of the pony truss bridges with these two variations. There are nearly twice as many bridges fitted with tapered web posts as compared to parallel web posts. This variation appears to be only the result of the selection by the buyer who opted for either parallel or tapered web posts for purely aesthetic reasons.

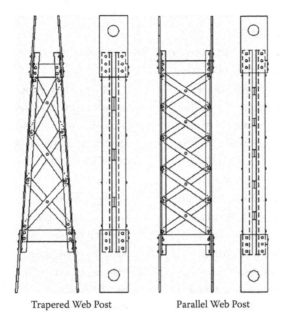

Trapered Web Post   Parallel Web Post

FIGURE 7.10   Vertical web post variations used on pony trusses.

END POST—LOWER CHORD CONNECTIONS

On pony truss bridges, the connection between the end post and the top chord was made using a special cast iron connector with the lower chord extending through and then held with a nut. A comparison of these two styles of end post-top chord connections is shown in Figure 7.11. Table 7.3 gives a summary of the remaining pony truss bridges with different end post connections for the upper and lower chords. The use of both end post configurations appears to be about equal. There does not appear to be any preference on connection type with age, that is both early and late bridges show both types of connections, and there is no trend with either span or number of panels. The choice on end post connection appears completely random at this time.

LOWER EYE BAR CHORDS

Wrought iron eye bars used as lower chords show the widest variation of any members used on both the pony truss and through trusses. The different size eye bars used are summarized in Table 7.4. All pony truss bridges and nearly all through truss bridges used only pairs of eye bars on the lower chords (the exceptions are the 5 spans of the Aiken St. Bridge in Lowell Mass which carry two lanes of traffic and pedestrian walkways; three spans of the So. Washington St. Bridge in Binghampton, N.Y., and the Bardwell's Ferry Bridge in Shelburne, Ma.). Figure 7.12 shows the relationship between the total cross sectional area of the lower chord eye bars and the span length for both Pony Truss and Through Truss Bridges. While there appears to be considerable scatter in these observations, a more detailed review is needed to isolate those bridges which also were constructed with either single or double walkways, which increase the dead load of the bridge.

TABLE 7.2
Summary of Pony Truss Bridges with Tapered and Parallel Web Posts

Tapered Web Posts	Parallel Web Posts
Main Street, Talcottville	Minortown Rd., Woodbury
Melrose Ave., E. Windsor	Washington Ave., Waterbury
W. Main Street, Stamford (2)	Sheffield Street, Waterbury
Oliver Street, Stamford	Sharon Station Rd., Sharon
Oregon Rd., Meriden	Golden Hill Rd., Lee
Ashland Mill, Jewett City	Fort River, Amherst
Hallville Mill, Preston	Dow Ave., Franconia
Pumpkin Hollow Rd.	Corinth Rd., Hadley
Gilbert, Rd., Warren	Water St., Homer (east)
Blackstone Bikeway	Water St., Homer (west)
North Canal, Lawrence	Wall St., Homer
Delage Rd., Franconia	Pine St., Homer
Cleveland Bridge, Mahwah	Niobi Junction
Rhule Rd, Malta	
Silk St., Newark Valley	
Long Lane Road, Steuben Co.	
Wallace	
S. High St., Duncannon	
Arkwright Mills	
Stillwater Rd., Smithfield	
Highgate Falls	
Randolph Rd., Bethel	

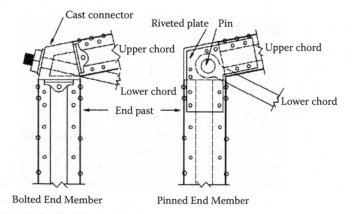

Bolted End Member   Pinned End Member

FIGURE 7.11   Variation in end post connections used on pony truss bridges.

TABLE 7.3

Summary of Different End Connections used on Pony Truss Bridges

Bolted Lower Chord End Connection	Pinned Lower Chord End Connection
Minortown Rd., Woodbury	Main Street, Talcottville
Melrose Ave., E. Windsor	Washington Ave., Waterbury
Hallville Mill, Preston	Sheffield Street, Waterbury
Ashland Mill, Jewett City	W. Main Street, Stamford (2)
Sharon Station Rd., Sharon	Oliver Street, Stamford
Pumpkin Hollow Rd.	Oregon Rd., Meriden
Gilbert, Rd., Warren	Golden Hill Rd., Lee
Delage Rd., Franconia	Fort River, Amherst
Dow Ave., Franconia	Blackstone Bikeway
Rhule Rd, Malta	North Canal, Lawrence
Corinth Rd., Hadley	Cleveland Bridge, Mahwah
Silk St., Newark Valley	Water St., Homer (east)
Long Lane Road, Steuben Co.	Water St., Homer (west)
Stillwater Rd., Smithfield	Wall St., Homer
Highgate Falls	Pine St., Homer
	Water St., Homer (east)
	Niobi Junction
	S. High St., Duncannon
	Randolph Rd., Bethel

SUMMARY

An inventory of the extant lenticular iron truss bridges manufactured by the Berlin Iron Bridge Company in East Berlin, Connecticut has been developed and is being used to provide some clues into the designs used by the Company. Comparisons of the geometries of the various components used at specific locations are being performed to provide a basis using "as-built' configurations. Only some initial comparisons have been made at this writing; more detailed evaluation is needed. Using the documentation developed by observations made on the extant bridges along with other documentation in HAER on bridges that have been lost, it is hoped that a more detailed picture of the design of these unique bridges will be possible.

As far as is known, there is no single complete list of lenticular bridges built by the Company; however, a partial list is currently being prepared by the author using a catalog of the Corrugated Metal Company (graciously provided by Mr. William Chamberlain), the list of bridges published by Darnell (1979), the author's personal copy of the Berlin Iron Bridge Company promotional catalogue circa 1890; various advertisements published in period trade journals and magazines; the author's personal collection of period postcards; the scant references available in the technical literature, and the list of extant bridges given in Table 7.1.

TABLE 7.4
Summary of Variations in Lower Chord Eye Bars

No. of Lower Eye Bars	Eye Bar Dimensions	Section Area (in.²)	Bridge
Pony Truss Bridges			
2	3⁄4 in. × 2 in.	1.5	Water St. (east)
2	3⁄4 in. × 2 1⁄2 in.	1.875	Minortown Rd. Melrose Ave. Corinth Rd.
2	1 in. × 2 in.	2	Fort River Wall St. Homer
2	1 1/16 in. × 2 in.	2.125	Water St. (west)
2	3⁄4 in. × 3 in.	2.25	Talcottville Oliver St.
2	1 1/2 in. × 1 1/2 in.	2.25	Bethel
2	1 in. × 2 1⁄2 in.	2.5	Washington Street Sheffield St. Hallville Mills Silk Street Ruhle Rd.
2	7/8 in. × 3 in.	2.625	Ashland Mill
2	1 1/8 in × 2 1⁄2 in.	2.8125	Pine St. Homer Smithfield
2	1 in. × 3 in.	3	Oregon Rd. Golden Hill Rd. Pumpkin Hollow Gilbert Rd. Blackstone Bikeway Highgate Falls
2	1 in. × 3 in. & 2 in. × 3 1⁄4 in.	36.5	Dow Ave.
2	3⁄4 in. × 4 in.	3	Delage Rd.
2	1 in. × 3 5/8 in.	3.625	Arkwright Mill
2	1 3/16 in. × 3 1/8 in.	3.71	Mahwah
2	1 3⁄4 in. × 3 in.	3.75	Duncannon
2	1 in. × 4 in.	4	W. Main St., Stamford North Canal, Lawrence
Through Truss Bridges			
2	3⁄4 in. × 2 1⁄2 in. & 1 in. × 2 in.	1.8752	Taylor Town Line
2	7/8 in. × 2 1⁄2 in.	2.1875	Kelsey Street
2	3⁄4 in. × 3 in.	2.25	Santa Cruz Ranch Main St. Moosup

TABLE 7.4 (continued)
Summary of Variations in Lower Chord Eye Bars

No. of Lower Eye Bars	Eye Bar Dimensions	Section Area (in.²)	Bridge
	Pony Truss Bridges (continued)		
2	1 in. × 3 in.	3	DelhiGalvin Rd. Chichester
4	1 in. × 3 in.	3	Bardwell's Ferry
2	1 1/8 in. × 3 in.	3.375	Pierceville
2	1 1/4 in. × 3 in.	3.75	Brunswick Ave.
2	1 in. × 4 in.	4	Ouaquaga
4	1 in. × 4 in.	4	South Washington St.
2	1 1/2 in. × 3 in.	4.5	Neshanic Station
2	1 1/8 in. × 4 in.	4.5	Boardman's Bridge
2	1 1/4 in. × 5 in.	6.25	Brackenridge Park Lover's Leap
4	1 5/8 in. × 4 in.	6.5	Aiken St.
2	1 3/8 in. × 5 in.	6.875	Highgate Falls

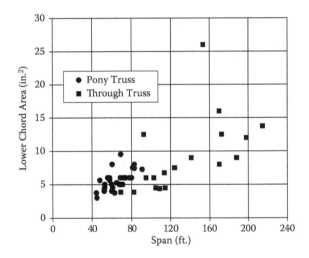

FIGURE 7.12 Relationship between lower chord area and span length.

REFERENCES

Boothby, T., 2004. Designing American Lenticular Truss Bridges 1878–1900, Industrial Archeology, Vol. 30, No. 1, pp. 5–17.

Campin, F., 1871. On the Application of Iron to the Construction of Bridges, Girders, Roofs and Other Works. Lockwood & Co.: London, 186 pp.

Darnell, V., 1979. Lenticular Bridges from East Berlin, Connecticut. Journal of the Society for Industrial Archeology, Vol. 5, pp. 19–32.

Douglas, W.O., 1877. Wrought Iron Bridge Designs. Scientific American Supplement, Vol. 4, No. 80.

DuBois, A.J., 1883. The Strains in Framed Structures. John Wiley & Sons: New York.

Guise, D., 2007. Development of the Lenticular Truss Bridge in America. Journal of Bridge Engineering, ASCE, Vol. 12, No. 1, pp. 120–129.

Kanonik, M.C., 2006. Rehabilitation of the Hadley Bow Bridge. Society for Industrial Archeology Newsletter, Vol. 35, No. 4, pp. 14–15.

Lutenegger, A.J. and Cerato, A.B., 2005. Lenticular Iron Truss Bridges in Massachusetts. Civil Engineering Practice, Journal of the Boston Society of Civil Engineers, Vol. 20, No. 1, pp. 53–72.

Matheson, E., 1873. Works in Iron: Bridge and Roof Structures. E. & F. N. Spon: London.

Merriman, M. and Jacoby, H.S., 1906; 1920. A Text-Book on Roofs and Bridges: Part I—Stresses in Simple Trusses. John Wiley & Sons Inc.: New York.

Rankine, W.J.M., 1874. A Manual of Civil Engineering. Charles Griffin & Co.: London.

Ritter, A., 1879 (translated by H.R. Sankey). Elementary Theory and Calculation of Iron Bridges and Roofs. E. & F. N. Spon: London.

Shreve, S.H., 1873. A Treatise on the Strength of Bridges and Roofs. D. Van Nostrand Publishers: New York.

8 Wind and Truss Bridges
Overview of Research

Frederick R. Rutz and Kevin L. Rens

CONTENTS

INTRODUCTION

The inspiration for this paper finds its roots in Frank Hatfield's, "Engineering for Rehabilitation of Historic Metal Truss Bridges", presented at the 7th Historic Bridges Conference.[13] Dr. Hatfield suggested that conventional truss analysis may underestimate the strength of a metal truss bridge and he observed that today's requirements for design wind load are much more demanding than historical design practice.

The focus of recent research at the University of Colorado at Denver has been on the lateral stiffening effect of decks in historic truss bridges. There is strong evidence from gravity load tests that truss bridges are actually stronger than traditional analysis predicts.[1,5,6,7,11,15,20,22,24,33] This suggests that the deck, which is neglected by traditional analysis, stiffens the structure. If the deck stiffens the bridge under gravity loads (i.e. in the vertical direction) it is reasonable to consider the deck as stiffening the bridge under wind loads in the lateral direction too.

TABLE 8.1

Summary of Wind Pressure Recommendations from the Late 19th Century and Early 20th Century

Year, Author	Span 20m (60 ft.)	30m (100 ft.)	60m (200 ft.)	300m (1000 ft.)	450 (1500 ft.)
1881, Smith	1.44 kPa (30 psf)	1.44 kPa (30 psf)	1.44 kPa (30 psf)	1.44 kPa (30 psf)	—
1884, Waddell	—	1.92 kPa (40 psf)	1.68 kPa (35 psf)	1.44 kPa (30 psf)	1.44 kPa (30 psf)
1898, Waddell	1.92 kPa (40 psf)	1.92 kPa (40 psf)	—	—	1.20 kPa (25 psf)
1905, Cooper	2.40 kPa (50 psf)	1.44 kPa (30 psf)	1.44 kPa (30 psf)	—	—
1916, Waddell	1.68 kPa (35 psf)	1.68 kPa (35 psf)	1.68 kPa (35 psf)	1.20 kPa (25 psf)	—

With permission from ASCE.

BACKGROUND

All structures respond to the application of load by altered states of material stress in members and connections and by deformations. Deformations from gravity loads lead to downward deflections. In a like manner, lateral loads such as wind and earthquake induce deflections and material stresses in the lateral direction.

WIND LOADS

In the absence of formal standards, 19th century bridge engineers based the determination of wind loads on their own reasoning. The thinking of several prominent American engineers is summarized in Table 8.1.[30] The recommendations of Smith, Waddell, and Cooper appear to be based on their own conclusions, formed from years of experience in bridge engineering work.[9,35,38,39,40] While the thinking of all bridge engineers probably evolved over this period, that of Waddell is well documented in his three books, which provide insight into the reasoning behind bridge design criteria in his day.[38,39,40] For spans in the range of 30 meters to 60 meters (100 feet to 200 feet), typical for many surviving truss bridges today, the design wind pressure was on the order of 1.44 to 1.92 kPa (30 to 40 psf). This is significantly lower than the requirements of today's AASHTO *Guide Specifications for the Design of Pedestrian Bridges,*[2] which mandates 3.59 kPa (75 psf)—about double the original design value. For a discussion of evolution of design wind pressure, see References 29 and 31.

Thus today's engineer must contend with a significantly higher wind pressure than the bridge was designed for, a factor to be combined with traditional structural analysis methods that underestimate the lateral strength of the bridge. No wonder so many historic truss bridges cannot comply with modern requirements!

Traditional Model

Traditional structural analysis of a truss bridge is based on a "skeleton" frame analysis, the classic textbook method, which has been used since Squire Whipple published the method of joints in 1847.[42] While the *techniques* of analysis have changed—computer analysis vs. hand calculations—the *basis* has remained the same. The "computer" in the 19th century design office was the individual who performed the calculations, using the classic methods of joints and sections or perhaps graphical methods that simplified some of the arithmetic. Today's practitioner using one of the many readily available computer programs is really utilizing matrix algebra. The computer is now a machine, but it does the same job—it completes the calculations. While the techniques have changed, the fundamental basis is still the same—a skeleton structure is assumed.

As noted above, there are many examples of gravity load tests compared to analytical results based on the skeleton analyses, with the results demonstrating that vertical stiffness is actually greater than calculated. While there are instances, such as the Cornish-Windsor Covered Bridge[12] and a Pratt truss in Franklin County, OH,[34] of engineers having addressed the deck in the lateral analysis, little can be found in publications. This paper is offered to help fill a void in the literature by describing a method used to evaluate truss bridges under lateral loads.

Proposed Model

Because there is strong evidence that decks stiffen truss bridges, it follows that an analytical model that goes beyond the traditional skeleton to include the deck would predict a more realistic response that the traditional model alone. We decided to pursue this hypothesis to determine if the stiffening effect is significant. Now to our base premise: Why do physical bridges have a lesser response to load than traditional analysis predicts? Could it be that features such as decks stiffen the structure but are not customarily accounted for in structural analysis? It is to these questions that our research has been directed.

RESEARCH

Researchers at the University of Colorado at Denver focused on the following:

- Traditional analysis of existing truss bridges for wind pressure as stipulated by different building codes
- Analysis of existing bridges using AASHTO-mandated wind pressure using both traditional analysis and utilizing the finite element method (FEM) to account for the stiffening effect of decks
- Field testing of existing bridges under wind conditions
- Comparison of the analyses to the field data to verify analytical models
- Response of bridge with no deck
- Response of portals
- Development of a low-cost strain transducer to facilitate field testing

CODE STUDY

Five historic truss bridges, all pin-connected trusses that varied in span from 24-meters (80-feet) to 49-meters (160-feet) and built between 1886 and 1913, were evaluated.[8] Their conditions were observed and as-built dimensions obtained. Each was analyzed using RISA-3D,[26] a 3D structural analysis program, under both the gravity and wind load requirements of five different codes:

- *Uniform Building Code*, 1997 edition.[37]
- American Society of Civil Engineers, *Minimum Design Loads for Buildings and Other Structures*, 2002 edition.[3]
- American Railway Engineering Association, *Manual for Railway Engineering*, 1987 edition.[4]
- National Building Code of Canada, 1990 edition.[21]
- American Association of State Highway and Transportation Officials, *Guide Specification for Design for Pedestrian Bridges*.[2]

For each case the bridge was treated as if it were rehabilitated for pedestrian use. It was found that all the codes had similar requirements for live load, with only 15% variation among their requirements. On the other hand, there was a range of differences of over 300% among the code requirements for wind pressure.

Three-dimensional (3D) analyses revealed that all of the trusses were adequate under gravity loads for all of the codes. However, under wind loads, all of the bridges under all of the codes had overstressed members. Typical overstressed members were:

- End diagonal members of portal frames
- Cross-bracing in the plane of the bottom chords
- Bottom chord eyebars on the windward side at mid-span, which theoretically went into compression
- Bottom chord eyebars on the leeward side at the anchored bearing end, which also went into compression

LATERAL LOAD PATHS STUDY

Six surviving through-truss bridges, summarized in Table 8.2 and shown in Figures 8.1 through 8.10, were selected for detailed analysis and testing. Similarities among all six bridges include:

- Through-trusses
- Former highway bridges
- Pratt trusses or Pratt derivatives
- Pin-connected with moment-resisting portal frames at the ends
- Metal—either wrought iron or steel
- Horizontal trusses consisting of rod X-bracing intended to resist lateral loads

Differences among them include:

- Different deck types (and for one, no deck at all)
- Different spans

TABLE 8.2
Summary of Bridges

Bridge	Constructed	Truss Type	Span	Deck/Stringers
Fruita	1907	Parker	47 m. (155 ft.)	Timber planks on timber stringers.
Blue River	1895	Pratt	25m. (80 ft.)	Two layers of orthogonal timber planks on steel stringers.
Prowers	1909	Camelback	49m. (160 ft.)	Asphalt pavement on corrugated metal deck on steel stringers.
Rifle	1909	Pennsylvania	73m. (240 ft.)	Asphalt pavement on corrugated metal deck on steel stringers.
San Miguel	1886	Pratt	43m. (142 ft.)	Gravel road base on semi-circular corrugated metal pipe segments on steel stringers.
Four-Mile	1900	Pratt	36 m. (119 ft.)	No stringers. No deck.

- Different railings
- Different local damage/deterioration
- Different local topography (affects wind pressure distribution)
- Different bearing conditions

Models

Four analytical models, all 3D, were developed for each bridge. They are called:

- Skeleton model. This is the traditional method of analysis, although some engineers will still use a 2D representation of the trusses and combine the results with another 2D representation of the upper and lower horizontal trusses. (As a result of our studies, we recommend 3D analysis because it addresses lateral response much more reliably than the combined 2D approach). The wind pressure is the AASHTO-mandated 75 psf. An example is shown in Figure 8.11.
- Deck model. Includes the stringers as frame elements (similar to the skeleton model), but also includes the deck using finite element quadrilateral elements. The wind pressure is the AASHTO-mandated 75 psf. An example is shown in Figure 8.12.
- Diaphragm model. The stringers and deck are included. The deck is treated as a rigid diaphragm, that is, theoretically rigid in its own plane. This is unrealistic as it overstates the stiffening effect of the deck, but does represent an upper bound on lateral stiffness. The wind pressure is the AASHTO-mandated 75 psf. An example is shown in Figure 8.13.
- Verification model. Similar to the deck model, but modified to approximate the actual condition of the bearings (boundary conditions) at the time of field-testing and to approximate the actual wind pressure distribution that occurred during testing. The verification model is compared against the test results. A good correlation with test data suggests the deck model is accurate. An example is shown in Figure 8.14.

FIGURE 8.1 Undated photograph of Fruita Bridge over the Colorado River, near Fruita, Colorado. This three-span steel Parker truss was built in 1907. It has steel floor beams and timber stringers covered by a timber deck. The deck timbers in place over the timber stringers and the wood rail can be seen. The University of Colorado Denver instrumented the span in the foreground in 2004. (Photo courtesy of Museum of Western Colorado.)

FIGURE 8.2 Fruita Bridge deck. The deck timbers are spiked to timber stringers, which bear on steel floor beams. The deck is discontinuous in that there are gaps between the deck timbers. The short vertical timbers are the remnants of the former wood rail.

FIGURE 8.3 Blue River Bridge over the Blue River near Silverthorne/Dillon, Colorado. This steel Pratt truss has five bays, with a timber deck on steel stringers. It is believed to have been built approximately 1895 as the Two-Mile Bridge near Breckenridge, Colorado and moved to this site at a later, but unknown, date. Closed to vehicular use, it still serves as a pedestrian crossing of the Blue River. (With permission from ASCE.)

FIGURE 8.4 Blue River Bridge deck. The deck consists of longitudinal "running boards" on transverse timbers on steel stringers. The orthogonal criss-crossing of running boards and deck timbers creates a much more continuous deck than that at Fruita Bridge. The steel stringers bear on and are mechanically attached to the steel floor beams.

FIGURE 8.5 Prowers Bridge over the Arkansas River, near Lamar, Colorado. The Camelback Pratt through truss span, seen in this photo, was instrumented because it received the greatest wind exposure. It survived a major flood of the Arkansas River in 1921 and served as a highway bridge until its abandonment in 1994 when a nearby replacement bridge was constructed. (With permission from ASCE.)

FIGURE 8.6 Prowers Bridge deck. The steel stringers are riveted to the floor beams, suggesting that the stringers are original. However, the corrugated bridge deck is clearly a replacement. Asphalt pavement, now significantly weathered, was installed over the corrugated deck. (With permission from ASCE.)

FIGURE 8.7 Rifle Bridge over the Colorado River at Rifle, Colorado. This Pennsylvania truss has steel floor beams with steel stringers, covered by a corrugated metal deck with asphalt pavement. It has been abandoned since the late 1960's, when a replacement bridge was constructed. (With permission from ASCE.)

FIGURE 8.8 Rifle Bridge deck. The deck of (weathered) asphalt pavement on corrugated bridge deck on a combination of steel and timber stringers is similar to that of the Prowers Bridge.

FIGURE 8.9 Fifth Street Bridge over Grand River, Grand Junction, CO. Built in 1886, one span of this five-span, wrought iron Pratt truss was relocated to the San Miguel location in the 1930's. The original timber deck and very simple, single board railing can be seen. (Photo courtesy of Museum of Western Colorado.)

FIGURE 8.10 San Miguel Bridge deck. The deck, believed to have been installed in 1964, consists of gravel roadbase on semi-circular corrugated metal pipe segments, which bear on the bottom flanges of steel stringers. It now has a roadway of gravel on an unusual system of semi-circular lengths of corrugated metal pipe set between steel stringers. A steel vehicular rail has replaced the original railing. It served the mining industry in western Colorado until the 1980's. Abandoned since 1990, it remains the oldest bridge originally built in Colorado. (With permission from ASCE.)

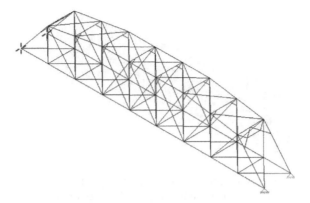

FIGURE 8.11   Skeleton model of Prowers Bridge. (With permission from ASCE.)

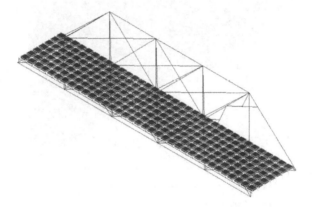

FIGURE 8.12   Deck model of Blue River Bridge. The deck consists of quadrilateral finite elements. Further, the deck elements are offset from stringer elements, and the stringer elements offset from floor beam elements illustrated in Figure 8.15. The offset dimensions are consistent with measured as-built dimensions. (With permission from ASCE.)

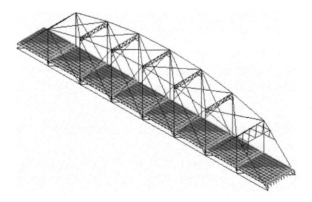

FIGURE 8.13   Diaphragm model of Fruita Bridge. The deck treated as a single diaphragm, shown here as part of each bay.

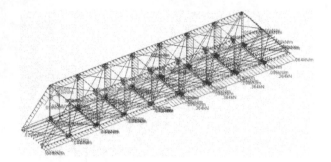

FIGURE 8.14 Verification model of San Miguel Bridge. The boundary conditions have been modified to simulate existing conditions and wind pressure varied to approximate the measured conditions. (With permission from ASCE.)

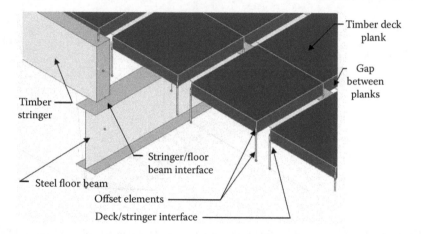

FIGURE 8.15 Detail of deck and stringer modeling. This computer-generated rendering of the timber deck planks on timber stringers on steel floor beams is for the Fruita Bridge deck, so gaps are present between the rows of plate elements. Continuous deck for other bridges had the corner nodes of deck elements connected. (With permission from ASCE.)

Discussion of Analytical Models

Bridges with both timber decks and replacement decks of corrugated metal with pavement were studied. A computer-generated rendering of an example of finite element modeling for a timber plank deck is shown in Figure 8.15.

Timber Decks

Two of the bridges had timber decks: Fruita Bridge and Blue River Bridge. However, the deck configurations were different.

At Fruita Bridge, as can be seen in Figure 8.2, the deck of individual timber planks spiked to the timber stringers leaves small gaps between the planks, thus this deck is discontinuous. Even with the gaps, the deck model indicated a stiffening effect from the stringers/deck construction. A measure of this effect is the calculated axial force in the bottom-chord eyebars. Despite the tension-inducing effect of the bridge's self-weight, the AASHTO-mandated wind load of 75 psf on the skeleton model resulted in net compression in the windward bottom-chord eyebars—a big problem for designers.

Because such compression would result in buckling of the affected members (which could lead to collapse of the structure) designers will take measures to preclude the compression. Such measures might include adding weight to the bridge to induce additional tension in the bottom chord eyebars, which reduces the remaining capacity of the structure for live load, or reinforcing those members, which alters the historic character of the structure to be preserved in the first place. These measures add expense, and expense can derail a preservation project too.

For the deck model, individual deck planks with gaps between the planks were approximated. The actual deck planks are spiked to the timber stringers so the model approximated the spiked connection as pinned. The deck elements were modeled to simulate the gaps between the actual deck planks, shown in Figure 8.15. The diaphragm model was then analyzed, although it is considered unrealistically stiff in the lateral direction because actual wood decks with gaps between the planks clearly do not possess in-plane rigidity. Still, the diaphragm model represents a theoretical upper bound on the lateral stiffening effect.

The other case of a timber deck is the bridge over the Blue River (Figure 8.4). While it has transverse deck planks similar to Fruita Bridge, longitudinal running boards have been added on top of the deck planks. The two layers of mutually orthotropic timbers, well spiked together, approximate a single solid deck. The behavior of interconnected deck elements is quite different from the "gapped" deck at Fruita Bridge. The deck was modeled with a grid of interconnected plate elements connected to the supporting stringers with rigid offset frame elements, shown in Figure 8.12.[14] One might expect this virtually solid deck to behave more closely to a rigid diaphragm than the deck at Fruita Bridge and this expectation was confirmed by the analyses. Both Fruita and Blue River Bridges were modeled using RISA 3D.[26]

Corrugated Metal Decks

For both Prowers Bridge and Rifle Bridge, the corrugated metal deck/pavement construction, shown in Figures 8.6 and 8.8, was modeled with interconnected elastic plate elements similar to Blue River Bridge. This is a simplification because the corrugated metal is much stiffer in the direction of the flutes, and more flexible in the direction transverse to the flutes. The "accordion effect" of the flutes results in a response sensitive to stress in the longitudinal direction, but not nearly so to stress in the transverse direction. The decks were modeled with a grid of elastic plate elements (similar to Figure 8.12) of constant stiffness in all directions as if the deck were an isotropic solid, this done because of software limitations. Values for material stiffness (modulus of elasticity, or E) were input for plate/shell elements and apply in all directions, thus a simple approach to a complex problem was adopted. Despite this limitation, we felt that a methodology utilizing readily available software tools would be more beneficial to preservation efforts than the use of more expensive software with greater analytical precision. Different stiffnesses were studied in the course of analysis, and the stiffness with the best fit to field-acquired test data was adopted. Prowers Bridge[18] was modeled using RISA 3D and Rifle Bridge[36] was modeled using RAM Advanse.[25]

San Miguel Bridge had been originally constructed with a timber deck on timber stringers with five spans as the Fifth Street Bridge over the Colorado River at Grand Junction in 1886, shown in Figure 8.9. When that bridge was replaced in the 1930's, one of the spans was relocated to the San Miguel River site. It was then subjected to heavy live loads from ore-carrying trucks in an active mining region of the Colorado Plateau. The current deck, consisting of gravel roadbase on semi-circular segments of corrugated metal pipe supported on steel stringers, was installed in 1964 (Figure 8.10). The thickness of the gravel roadbase produces a heavy dead load of approximately 3.54 kPa (74 psf), the highest of all the deck dead loads studied. This deck was modeled using RISA 3D and interconnected plate elements were used to represent the gravel roadbase, however the offset elements were modeled such that the deck elements were at the same elevation as the stringer top flanges.[10] This model representation is really no different than that used at Prowers, and Rifle, although the physical deck construction was quite different. As at Prowers and Rifle, this was done for modeling simplicity.

These techniques are described in greater detail in the report on a research project completed by the Department of Civil Engineering at the University of Colorado at Denver for the National Center for Preservation Technology Transfer[23] and in the individual theses.[10,14,18,27,36]

Tests

How accurate is the deck model? To verify each model, a test on each bridge was carried out. At each bridge selected members were instrumented to measure strain under conditions of wind from a direction transverse to the bridge. Instrumentation was set up at the bridges and data was collected during windy conditions. The instrumentation system was designed to obtain measurements of wind speed, wind direction, and strains from selected members all at a rapid sampling interval. Anemometers and a wind direction sensor were used to obtain wind speed and direction data at multiple locations virtually simultaneously. Strain data was obtained at sixteen different locations, selected for anticipated high axial forces or moments under to lateral load. Figures 8.16 and 8.17 show the location of instruments for a typical bridge. Some strain transducer installations are shown in Figure 8.18. Figure 8.19 illustrates typical strain measurements, with changes in strain correlating with changes in wind velocity. See References 23 and 32 for detailed descriptions of the instrumentation system.

Comparisons

The measured force vs. calculated force for both the skeleton model and the deck model for each bridge is presented in Table 8.3.

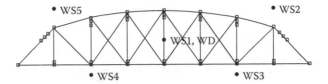

FIGURE 8.16 Diagram of Fruita Bridge, illustrating the locations of anemometers (WS1 – WS5) and wind direction sensor (WD). North is to the left. (With permission from ASCE.)

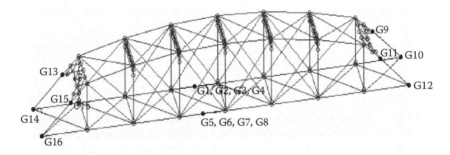

FIGURE 8.17 Diagram of Fruita Bridge, illustrating the locations of the strain transducers. North is to the left. The wind direction was from the west, orthogonal to the bridge. (With permission from ASCE.)

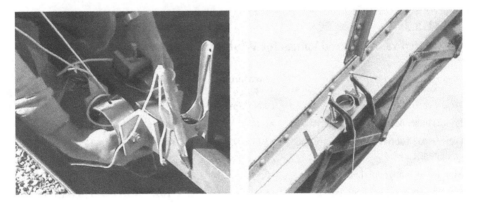

FIGURE 8.18 Strain transducers being installed on bottom-chord eyebars at Prowers Bridge and strain transducer installed on an end diagonal of a portal at Fruita Bridge. Each component has a lanyard attached to it, as a precaution against falling into the river below.

Model and Test Verification of Bridge with No Deck

If tests have verified the stiffening effect of decks predicted by analysis, then we would expect a bridge that has no deck to behave in a manner similar to that predicted by a skeleton analysis. To confirm this assumption, a bridge with no deck was instrumented and tested, and the results compared to that from a skeleton model. When the former Elk River Bridge, built in 1900 near Steamboat Springs, CO, was replaced in 1990, it's deck was removed and it's frame was relocated to the edge of a

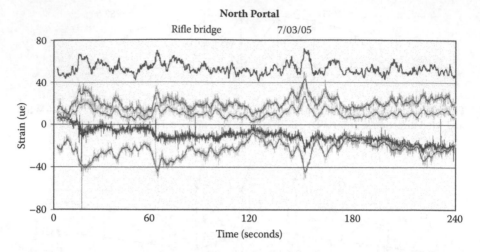

FIGURE 8.19 Example of strain measurements. The upper trace is the wind speed, averaged from seven anemometer measurements, and presented at an arbitrary scale for presentation purposes. The four lower traces are strain measurements from different members. This example is from the north portal of Rifle Bridge. Note that changes in strain correspond to changes in wind velocity. From such measurements, forces in members were determined for comparison with forces determined from the verification models. (With permission from ASCE.)

TABLE 8.3
Calculated vs. Measured Values for Windward Bottom Chord

Bridge	Measured Force kN (kips)	Calculated Force Skeleton Model kN (kips) Difference %	Calculated Force Deck Model kN (kips) Difference %
FruitaTimber deck	−1.44 (−0.32)	−2.07 (≠0.46) 44%	−1.37 (−0.31) 5%
ProwersAsphalt pavement on corrugated metal decking	−3.11 (−0.70)	−4.09 (−0.91) 32%	−2.71 (−0.61) 13%
Blue RiverOrthogonal timber deck	−0.54 (−0.12)	−0.54 (−0.12) 0%	−0.44 (−0.1) 19%
San MiguelGravel roadbase on CMP segments	−6.89 (−1.55)	N/A	−5.25 (−1.18) 24%
RifleAsphalt pavement on corrugated metal decking	−8.41 (−1.89)	−12.37 (−2.75) 47%	−7.34 (−1.65) 13%

wheat field on top of a small hill. Because the skeleton structure did not have much surface area, a sail was set up to intercept wind, shown in Figure 8.20.[28] The measured response compared favorably (within 8%) to that from a 3D skeleton model.

Portal Modeling Question

Modeling of boundary conditions for portal frames could be treated as either pinned or fixed, illustrated in Figure 8.21. 19th century bridge engineers assumed base

FIGURE 8.20   Overview of sail in place during data collection at Four Mile Bridge.

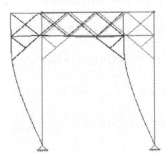
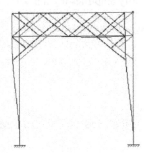

Portal frame with pinned ends   Portal frame with fixed ends

FIGURE 8.21   Portal frame modeled with fixed ends is shown at the right. The same portal frame modeled with a release from rotation is shown at the left. The fixed end case was typically used prior to 1916. Design based on pinned ends appears to have been first introduced in 1908. (With permission from ASCE.)

fixity.[41] It was not until the early 20th century that the assumption of rotational releases ("pinned" about the bridge longitudinal axis) began to be used.[19] Typical practice today is to assume rotational releases at the base, which leads to higher calculated flexural stresses in the end diagonal members than the members were initially designed for. However, testing revealed both the tops and bottoms of the portals responded to wind more or less equally, suggesting base fixity, at least for the measured wind speeds. This response occurred in all six cases.

Conclusions

These studies suggest:

- The forces in the bottom chord eye bars are lower than traditional skeleton analysis would indicate, attributed to the stiffening effect of the deck in the lateral direction.

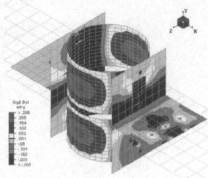

FIGURE 8.22   Strain transducer and its finite element stress analysis results. The transducer consists of a 3" diameter steel ring, bolted to steel angles. The strain gage is on the inside ring surface, 90 degrees from the axis of the bolts. The transducer senses axial strain and mechanically amplifies it via flexure of the ring. (With permission from ASCE.)

- To take advantage of this, the deck can be included in the structural analysis using readily-available software.
- This methodology has been verified by tests on real bridges in real wind conditions.
- Portals behave as if the bases are fixed, at least for the wind velocities measured.
- Economical, reusable strain transducers can be used for testing.

TRANSDUCER DEVELOPMENT

As all researchers engaged in installation of strain gauges on outdoor structures well know, such activities are at best labor-intensive, involving surface preparation work of grinding to remove paint and corrosion, sanding to create a smooth uniform surface, polishing, the delicate operation of bonding and clamping the strain gage, and soldering lead wires. To avoid these difficulties, we developed a reusable, modular strain transducer.

Each strain transducer consisted of a steel ring with a strain gage adhered to the inside surface as shown in Figure 8.22. Strain gages were applied at a location of high stress on the ring. The ring was attached to two steel angles, which were used for clamping the transducer to the member being studied. Fundamentally, the transducers sense axial strain, which is amplified by flexural deformation of the ring. The true strain in the member under study is obtained by multiplying the transducer strain by a predetermined factor, which had been determined theoretically and confirmed experimentally by laboratory tests (Figure 8.23).[16,17]

NOTES ON WIND OBSERVATIONS

In a commentary on the design criteria for the Firth of Forth Bridge in Scotland, American engineer Theodore Cooper expressed his belief that the 2.68 kPa (56 psf) design lateral pressure imposed upon the designer of the Forth Bridge, was an

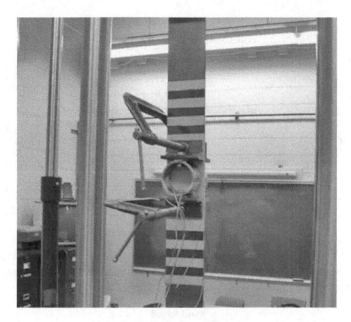

FIGURE 8.23 Laboratory test of ring transducer clamped to a steel plate, which simulates an eyebar of a truss bridge. (With permission from ASCE.)

"absurd requirement."[9] While Cooper agreed that it might be conceivable for a pressure of 2.68 kPa (56 psf) to strike a relatively short length of the bridge—say 30 meters (100 feet), he thought it impossible for such a high wind pressure to simultaneously load the total span of 521 meters (1,710 feet). Cooper's opinions bring us to consider some fundamental questions about the nature of wind: What is the size of a gust and how long does one last?

Our measurements brought some observations to light. Consider Figure 8.24, which illustrates simultaneous wind speed measurements from five different locations on Fruita Bridge. We observe that the wind speed, while following a trend suggested by the bandwidth of all measurements, is not at all identical at the several locations. Further, while we might expect the anemometers located at the highest elevations to show the greatest wind speeds and those at the lowest elevations the least wind speeds, at times we found the reverse to be true. These same conclusions can be drawn from Figure 8.25, which shows a much larger bandwidth of velocities, indicating even greater variations, at Prowers Bridge. Thus, based on the measurements we made, we must concur with Cooper in that wind pressure is not constant over the length of even these relatively short span bridges.

ACKNOWLEDGMENTS

The University of Colorado at Denver has conducted this work funded in part by Grant # MT-2210-04-NC-12 from the National Center for Preservation Technology Transfer and Grant # 2004-M1-019 from the State Historical Fund of the Colorado Historical Society. Further, the cooperation of the City of Fruita, CO, Summit

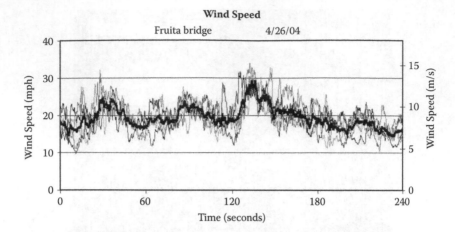

FIGURE 8.24 Wind speed as measured by five anemometers at Fruita Bridge. The bold line shows the average of all five anemometers. Note the variation band from the average velocity is relatively small. (With permission from ASCE.)

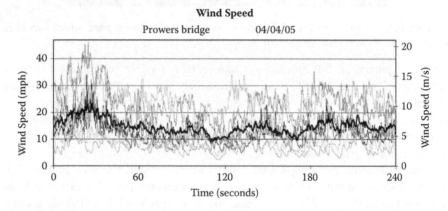

FIGURE 8.25 Wind speed as measured by seven anemometers at Prowers Bridge. The bold line shows the average of all anemometers. The wind velocity variation band is relatively large. (With permission from ASCE.)

County, CO, Bent County, CO, Montrose County, CO Garfield County, CO, and Selby Farms, Steamboat Springs, CO is gratefully acknowledged. The research efforts of graduate students Veronica Jacobson, Shohreh Hamedian, Kazwan Elias, and William Swigert resulted in major contributions to the project. Campbell Scientific, Logan, Utah, provided very good technical support with programming and operation of instrumentation.

REFERENCES

1. Aktan, A.E., Lee, K.L., Naghavi, R., and Hebbar, K., "Destructive testing of two 80 year-old truss bridges", *Transportation Research Record No. 1460*, Transportation Research Board, Washington, D.C., 1994, 62–72.

2. American Association of State Highway and Transportation Officials (AASHTO), *Guide Specifications for Design of Pedestrian Bridges*, Washington, D.C., 1997.
3. American Society of Civil Engineers (ASCE), *ASCE 7-02 Minimum design loads for buildings and other structures*, Reston, VA, 2002, Table C6-1.
4. American Railway Engineering Association (AREA), *Manual for railway engineering*, Washington, D.C., 1987.
5. Azizinamini, A., Mehrabi, A., Lotfi, H., and Mans, P., *Evaluation and retrofitting of historic steel truss bridges*, University of Nebraska – Lincoln, Nebraska, Department of Roads Research Project Number STB-STWD (13), Lincoln, 1997, 1–185.
6. Bakht, B., and Csagoly, P.F., "Testing of the Perley Bridge", Ontario Ministry of Transportation and Communication, Research and Development Division, Downsview, Ontario,1977, 1–3, 13, 28–36.
7. Bakht, B., and Jaeger, L.G., "Bridge testing—a surprise every time." *Journal of Structural Engineering*, American Society of Civil Engineers, Reston VA, 1990, 116 (5), 1370–1379.
8. Carroll, D., *Analysis of historic pin connected truss bridges for conversion to pedestrian use*, Dept. of Civil Engineering, University of Colorado at Denver, Denver, CO, 2003.
9. Cooper, T., "What Wind Pressure Should be Assumed in the Design of Long Bridge Spans?", *Engineering News*, 5 Jan. 1905, 15–16.
10. Elias, K., *Wind analysis of the historic San Miguel Bridge*, MS thesis, Univ. of Colorado at Denver, Denver, CO, 2007.
11. Elleby, H.A., Sanders, W.W., Klaiber, F.W., and Reeves, M.D., "Service load and fatigue tests on truss bridges." *Journal of the Structural Division*, American Society of Civil Engineers, Reston, VA, 1976. 2285–2300.
12. Fischetti, D., "Conservation Case Study for the Cornish-Windsor Covered Bridge," *APT Bulletin*, Albany, NY, 1991, Vol. 23, No.1, 22–28.
13. Hatfield, F. J., "Engineering for Rehabilitation of Historic Metal Truss Bridges", *Proceedings of the 7th Historic Bridges Conference*, Sponsored by Wilbur J. and Sara Ruth Watson Bridge Book Collection, Cleveland State University, Sept. 19–22, Cleveland, OH, 2001, 7–11.
14. Hamedian, S., *Analysis and testing of the historic Blue River Bridge*, MS thesis, Univ. of Colorado at Denver, Denver, CO, 2006.
15. Heins, C.P., Fout, W.S., and Wilkison, R.Y., "Replacement or repair of old truss bridges." *Transportation Research Record 607*, Transportation Research Board, Washington, DC, 1976, 63–65.
16. Herrero, T., *Development of strain transducer prototype for use in field determination of bridge truss member forces*, Civil Engineering Dept., Univ. of Colorado at Denver, Denver, CO, 2003.
17. Herrero, Rens, and Rutz, "Wind pressure and strain measurements on bridges. II: strain transducer development," *Journal of Performance of Constructed Facilities, American Society of Civil Engineers*, Reston VA, 2008, 22, 1, 12–23.
18. Jacobson, V. R., *Analytical techniques and field verification method for wind loading analysis of the historic Prowers Bridge*, MS thesis, Univ. of Colorado at Denver, Denver, CO, 2006.
19. Ketchum, M.S., *The design of highway bridges and the calculation of stresses in bridge trusses*, McGraw-Hill, New York, 1908, 141–155.
20. Nagavi, R.S., and Aktan, A.E., "Nonlinear behavior of heavy class steel truss bridges." *Journal of Structural Engineering*, American Society of Civil Engineers, Reston VA, 2003, 1114–1116.
21. NBC, *National building code of Canada*, Canadian Commission on Buildings and Fire Codes, National Research Council of Canada, Ottawa, 1990.

22. NCHRP 234, "Manual for bridge rating through load testing." *NCHRP Research Results Digest No. 234*, Transportation Research Board, National Research Council, Washington, DC, 1998, 97–99.
23. NCPTT 2004-25, *Load Paths in Historic Truss Bridges*, prepared by Dept. of Civil Engineering, University of Colorado Denver for National Center for Preservation Technology Transfer, Natchitoches, LA 2004.
24. Pullaro, J., "Rehabilitation of two 1890's metal truss bridges", *International Engineering History and Heritage*, ASCE, Reston VA, 2001, 215.
25. RAM, RAM Advanse, ver. 7.0, RAM International, Carlsbad, CA, 2005.
26. RISA, *RISA-3D*, ver. 4.5, Risa Technologies, Foothill Ranch, CA, 2002.
27. Rutz, F.R., *Lateral load paths in historic truss bridges*, PhD Thesis, Civil Engineering, Univ. of Colorado at Denver, Denver, CO, 2004, 181–227.
28. Rutz, F.R., *Lateral load paths in historic truss bridges*, PhD Thesis, Civil Engineering, Univ. of Colorado at Denver, Denver, CO, 2004, 212–227.
29. Rutz, F.R., *Lateral load paths in historic truss bridges*, PhD Thesis, Civil Engineering, Univ. of Colorado at Denver, Denver, CO, 2004, 116–127.
30. Rutz and Rens, "Wind loads for 19th century bridges: design evolution, historic failures, and modern preservation", *Journal of Performance of Constructed Facilities, American Society of Civil Engineers*, Reston, VA, 2007, 21, 2, 161.
31. Rutz and Rens, "Wind loads for 19th century bridges: design evolution, historic failures, and modern preservation", *Journal of Performance of Constructed Facilities, American Society of Civil Engineers*, Reston, VA, 2007, 21, 2, 162–163.
32. Rutz and Rens, "Wind pressure and strain measurements on bridges. I: Instrumentation/data collection system," *Journal of Performance of Constructed Facilities, American Society of Civil Engineers*, Reston, VA, 2008, 22, 1, 2–11.
33. Seyednaghavi, M., *Strength evaluation of the existing aged steel bridges based on destructive tests*. PhD Thesis, Civil and Environmental Engineering, University of Cincinnati, Cincinnati, OH, 1997, 19–23, 239–353.
34. Seyednaghavi, M., *Strength evaluation of the existing aged steel bridges based on destructive tests*. PhD Thesis, Civil and Environmental Engineering, University of Cincinnati, Cincinnati, OH, 1997, 124–125.
35. Smith, C.S., "Wind Pressure Upon Bridges", *Engineering News*, 1 Oct. 1881, 395.
36. Swigert, W.B., *Wind load analysis of a truss bridge at Rifle, Colorado*, M.S. thesis, Univ. of Colorado at Denver, Denver, 2007.
37. UBC, *Uniform building code*, International Conference of Building Officials, Whittier, CA, 1997.
38. Waddell, J.A.L., *The designing of ordinary iron highway bridges*, John Wiley & Sons, New York, 1884, 6.
39. Waddell, J.A.L., *De pontibus*, John Wiley & Sons, New York, 1898, 224 and Plate VIII.
40. Waddell, J.A.L., *Bridge engineering*, John Wiley & Sons, New York, 1916, 149–154.
41. Waddell, J.A.L, *Bridge engineering*, John Wiley & Sons, New York, 1916, 294.
42. Whipple, S., *A work on bridge building: consisting of two essays, the one elementary and general, the other giving original plans, and practical details for iron and wooden bridges*, H.H. Curtis, Printer, Utica, NY, 1847.

9 Mechanical Properties of Wrought Iron from Penns Creek Bridge (1886)

Stephen Buonopane and Sean Kelton

CONTENTS

INTRODUCTION

Penns Creek Bridge was a 93 foot long Pratt, through truss originally constructed in 1886 over Penns Creek in Penn Township, Centre County, Pennsylvania (Figure 9.1). The primary structural members of the trusses were constructed of wrought iron, with cast iron joints at the ends of the upper chord and at the supports. The design and construction have been attributed to the Columbia Bridge Works of Dayton, Ohio based on several unique features of the truss. In 1999 Penns Creek Bridge was determined to be National Register eligible as a locally significant example of a Pratt, through truss with all of its primary structural elements in unaltered condition.[8,17] The bridge was offered for sale and relocation for several years, but no suitable preservation plans came to fruition. In early June 2007 Penns Creek bridge was removed from service, and it has since been replaced with a prestressed concrete beam bridge.

Section 106 of the 1966 National Historic Preservation Act requires mitigation to offset adverse effects on National Register eligible properties. As a part of such mitigation efforts, the Pennsylvania Department of Transportation (PennDOT) Cultural

FIGURE 9.1   Penns Creek Bridge, May 2007.

Resources Management program, in cooperation with the Design and Construction divisions, arranged for material specimens salvaged from the bridge to be provided to Bucknell University for testing. The provisions for salvage of the materials were specified in the special conditions section of the contract for the demolition of Penns Creek Bridge and construction of its replacement.

Many 19th century truss bridges remain in service in the United States; however, their numbers are continually decreasing as they are replaced by bridges built to modern design standards. The evaluation and preservation of historic truss bridges remain a difficult challenge for present-day engineers and transportation authorities, due to their unfamiliar structural systems and archaic materials.[9] Rehabilitating such bridges requires specialized knowledge and skills across all levels of the evaluation, design and repair process.

Wrought iron has not been used as a primary structural material in the United States for over a century.[12,16] Many remaining wrought iron structures, such as Penns Creek Bridge, have historic engineering significance beyond their utilitarian function and are in danger of being demolished. Two factors that contribute to the disappearance of wrought iron structures are uncertainty about the behavior and properties of wrought iron and inherent variability of its mechanical properties.

BRIDGE DESCRIPTION AND CONDITION

Each Pratt truss of Penns Creek Bridge was symmetric about its centerline and was divided into 6 equal panels of 15.5 ft each (Figures 9.1 and 9.2). The lower chord was composed of two parallel plates, each 6" high by 3/8" thick. The plates were continuous across two panels, with bolted lap splices at panel points L2 and L4 (Figure 9.3). This style of lower chord is characteristic of the work of D. H. Morrison and the Columbia Bridge Works, in contrast to the more common eyebars and pins.[8,17]

The upper chord and inclined end posts were built-up sections consisting of a 6-3/8" deep I-shape, positioned horizontally, and riveted to two 5-7/8" deep channels (Figure 9.2). The joints at the ends of the upper chord (U1, U5) and at the supports

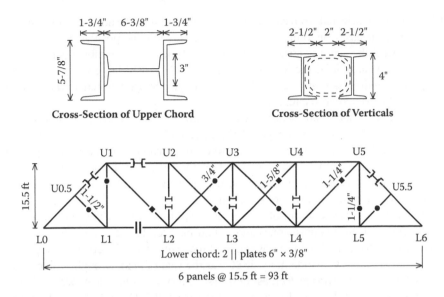

Cross-Section of Upper Chord   Cross-Section of Verticals

FIGURE 9.2   Truss elevation and member cross-sections.

FIGURE 9.3   Connection at lower chord panel point L4 (L2 similar).

(L0, L6) were fabricated from cast iron. The connections of the adjacent compression members to the cast iron joints were by bearing only with no positive connection.

In general the primary structural members of the trusses exhibited surface corrosion but no substantial section loss. Some secondary (and largely non-structural) components with small thicknesses did have enough corrosion to render them essentially non-functional or to cause failure.

The upper chord experienced severe distortion and distress as a result of rust occurring between the flanges of the I-section and the webs of the channels (Figure 9.4). The horizontal distance between the channels increased by approximately 1 inch due to the expansive forces of the rust. The distortion also caused severe local bending of the channel web in the vicinity of each rivet, and in some locations, longitudinal splitting of the channel web.

The manufacturer's stamp of "Jones & Laughlins" was visible on the web of the I-sections of the end post. Jones and Laughlins Ltd., located on the banks of the Monongahela River in Pittsburgh, was one of the primary producers of wrought iron in the United States. Coincidentally, 1886 was the year of peak wrought iron production at Jones & Laughlins as well as their first year of Bessemer steel production.[15]

The three central vertical members of the truss (L2-U2, L3-U3, L4-U4) were built-up sections composed of two 4" deep I-sections with short lengths of rectangular tube as a spacing member. These verticals were retrofitted circa 1996 with welded square batten plates.[8,17] These verticals remain in compression under service loads and have bearing-only connections at the top and bottom (Figure 9.3). Small tabs, providing no tension capacity, extended from the verticals into the bolted connection of the lower chord, presumably for alignment and ease of construction. Several of these tabs were found to be severely corroded.

The primary tension diagonals (U1-L2, U2-L3, L3-U4, L4-U5) were square rods of 1-1/4" or 1-5/8" dimension. The diagonals in the end panels were 1-1/2" diameter rod, and the counter-diagonals (L2-U3, U3-L4) were 3/4" diameter rod. All of the square and round diagonals have forged loops at the ends. Connections at the panel points were made through the use of additional pin plates (e.g. Figure 9.3), another identifying feature of Morrison and the Columbia Bridge Works.[17] The 3/4" diameter diagonal rods were presumably used during erection, as the self-weight of the bridge and applied traffic loads would have caused compression in these diagonals. Several of the pin plates connecting these diagonals were found to be severely corroded with substantial cross-section loss.

The transverse floor beams were constructed from plate and angles with varying floor depth. The bridge included diagonal X-bracing below the floor and in the top plane of the truss.

MATERIAL STRUCTURE OF WROUGHT IRON

The micro-structure of wrought iron consists of a ferrite matrix with slag inclusions distributed throughout. Slag is a non-metallic substance consisting primarily of oxides of iron and silica, and to a lesser extent of phosphorous and manganese. Slag is a result of the manufacturing process and its components come from the silicon naturally found in iron-ore and from the sand lining of iron furnaces. The majority

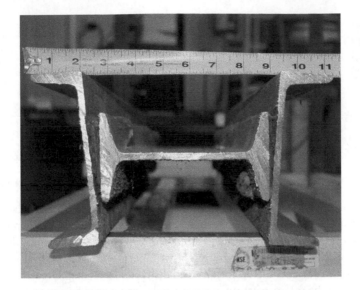

(a)

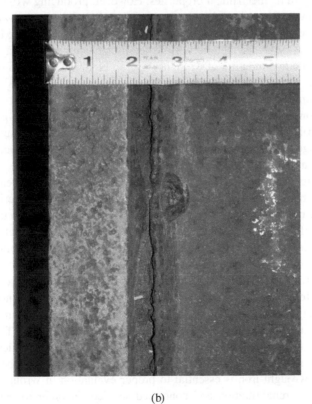

(b)

FIGURE 9.4 Photographs of upper chord showing distortion caused by pack rust: (a) cross section; (b) top view.

FIGURE 9.5 Large slag inclusion in specimen from the flange of an upper chord channel.

of the carbon content of pig iron is oxidized to carbon monoxide during the puddling process, but the remaining carbon may influence the behavior of wrought iron depending on the final state of the carbon. The working done during the puddling and piling process as well as the rolling of wrought iron into structural shapes (e.g. plates, bars, etc.) affects the uniformity and distribution of slag. Greater homogeneity achieved in the microstructure of the wrought iron would be expected to produce less variability in its mechanical properties. However, producing wrought iron with a consistent level of homogeneity required great skill and was not always achieved. Further, different grades of wrought iron were produced based on the intended use; e.g. railway rails vs. bridge members. More detail on the material structure, chemical composition and manufacture of wrought iron can be found in contemporary works such as Withey and Aston,[20] or modern studies such as Gordon,[11,12] Gordon and Knopf[14] or Kemp.[16]

Fabrication of the specimens revealed many large slag inclusions parallel to the direction of rolling and visible to the naked eye. One of the largest such inclusions from the flange of a channel section, is shown in Figure 9.5 and measured about 4 inches long and up to 1/4 inch wide.

Wrought iron's ability to function well as a structural material depends on attaining a balance between strength and ductility.[14] The strength of a material is characterized by the maximum stress that it can sustain in a uniaxial tension test. Ductility is a measure of the ability to plastically deform before failure, characterized by the elongation or reduction in cross-sectional area from a tension test. Toughness is the energy required to cause fracture, and it accounts for both the strength and ductility of material. Toughness is typically measured by the area under the stress-strain curve from a tensile test, or the impact energy required to fracture a notched specimen. Extremely strong iron may be associated with poor toughness, making it susceptible to brittle fracture and inappropriate as a structural material. The process of making wrought iron makes it very difficult to control these mechanical properties in a consistent manner, making them quite variable.[11] Understanding the engineering properties of wrought-iron is essential to proper evaluation of wrought iron structures for repair, rehabilitation and continued service. In order to obtain a better understanding of the engineering characteristics of wrought iron, three types of tests were conducted on the material salvaged from Penns Creek Bridge—tension, hardness and impact.

TABLE 9.1
Specimens, Member Types and Locations

Specimen Group	Description	Truss Members
UCA	Upper chord channel	L0-U1, U1-U5, U5-L6
UCB	Upper chord channel	
UI	Upper chord I-section	
VIA	Vertical I-section	L2-U2, L3-U3, L4-U4
VIB	Vertical I-section	
LPA	Lower chord tension plate	L0-L2, L2-L4, L4-L6
LPB	Lower chord tension plate	
RA	3/4" diameter rod	L2-U3, L4-U3
RB	1-1/4" diameter rod	L1-U1, L5-U5
RC	1-1/2" diameter rod	L1-U0.5, L5-U5.5
SA	1-1/4" square bar	L3-U2, L3-U4
SB	1-5/8" square bar	L2-U1, L4-U5

MECHANICAL TESTING METHODS

Table 9.1 summarizes the sources of all of the test specimens. Three tension tests were performed for each member type sampled and all tension specimens were taken along the direction of rolling, parallel to the fiber orientation. Tension specimens were machined in two sizes to accommodate different member geometries. Round specimens of 1/4 inch reduced diameter were machined from rods and square bars (specimen groups RA, RB, RC, SA and SB). Rectangular specimens of width 1/4 inch and thickness 1/4 inch were machined from the tension plates (LPA, LPB). The specimens from the channels (UCA, UCB) and I-sections (UI, VIA, VIB) were taken from the webs and had a width of 1/4 inch and a thickness of 0.15 inch. A gauge length of 1 inch was used for all specimens. Tension tests were performed in an Instron testing machine with a rate of elongation of 0.1 inch per minute was used following ASTM E8.[3]

Hardness tests were conducted with both Brinell and Rockwell B scales. The Brinell hardness test uses an indentation sphere of 10 mm diameter with a test weight of 3000 kgf; while the Rockwell B test uses a sphere of 1/16 inch diameter and a minor load of 10 kgf and a major load of 100 kgf. Hardness tests were conducted on a machined surface, with a minimum thickness in accordance with ASTM E10[4] and ASTM E18.[5] Three indentations were conducted on each sample.

Charpy V-notch impact specimens were machined and tested according to ASTM E23.[6] For round rods, square bars and tension plates, standard size specimens (10 mm by 10 mm) were used. For the channels and I-sections, reduced size specimens (5 mm thick by 10mm high) were machined from the flanges. Three specimens were machined from each member type and all specimens were oriented parallel to

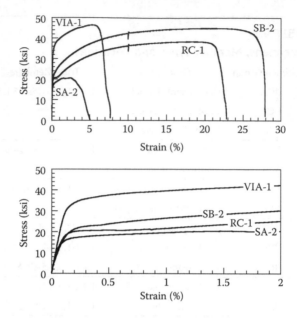

FIGURE 9.6 Representative stress-strain curves from tension tests.

the direction of rolling. Testing was performed using a Tinius Olsen impact testing machine, and results were measured in ft-lbs.

TEST RESULTS AND DISCUSSION

TENSION TESTS

Many of the stress-strain plots displayed the common characteristics of a ductile metal—elastic region, yielding, plastic region, strain hardening, and finally necking and failure. For certain specimens, however, the stress-strain behavior did not exhibit a sharp yield point and the stress continued to increase gradually through the plastic region. Four stress-strain curves representative of the variation among the tests can be found in Figure 9.6.

Table 9.2 summarizes the results of the mechanical properties derived from the tension tests—yield strength, ultimate strength, percent reduction in area and percent elongation. Three specimens were taken from each member, as listed in Table 9.1. The average property of the three specimens was considered to be a representative member value. The values reported in Table 9.2 are the minimum, maximum or average of the representative member values across all of the different members.

Figure 9.7 shows individual test data for yield and ultimate strengths with the horizontal axis grouped by member type and arranged approximately in order of increasing characteristic member thickness (web thickness for C or I shapes, thickness for flat plates, diameter for round rods, width for square bars). As expected for wrought iron, all of the mechanical properties exhibited a significant amount variation.

Withey and Aston[20] reports typical yield strengths of 40 ksi for 3/8 in diameter rods to 23 ksi for 2 in diameter rods, and tensile strengths (parallel to grain) of

TABLE 9.2
Summary of Average Test Results

Property	Min Value	Max Value	Average
Yield strength (ksi)	18.1	35.0	24.5
Ultimate strength (ksi)	26.1	48.8	39.2
% Reduction in area	11	41	25
% Elongation	7	29	20
Brinell hardness (BHN)	111	147	127
Rockwell B hardness	55	74	65
Impact strength (ft-lbs)	16	29	20

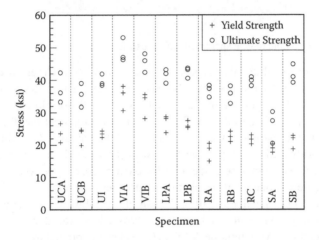

FIGURE 9.7   Individual test results for yield and ultimate strengths.

45 to 55 ksi across the same specimen size range. Data reported in Gordon[11] and Gordon and Knopf[14] from historical and modern tests on samples from numerous sources have a range of yield strengths from 32 to 50 ksi and ultimate strengths from 38 to 66 ksi. Data reported in Elban et al.[10] from three tensile tests on a circa 1857 I-section have yield strengths of 34, 39, and 42 ksi and ultimate strengths of 48, 54 and 48 ksi. Bright[7] reports data from a total of 516 tension tests parallel to the grain (compiled from historical and modern tests), resulting in a mean yield strength of 36 ksi with a standard deviation of 4 ksi (11% coefficient of variation), and a minimum credible value of 27.5 ksi.

AASHTO[1] Paragraph 6.6.2.3 specifies a minimum yield of 26 ksi and an ultimate of 48 ksi where no specific test data is available. In evaluating the tension strength of existing bridge members using AASHTO Paragraph 6.4.2,[1] the nominal yield and ultimate strengths are used in combination with LRFD resistance factors (ϕ) to calculate the member capacity. The values of ϕ specified by AASHTO[1] are based on assumed statistical properties of the yield and ultimate tensile strengths (mean, coefficient of variation and bias). The average values of yield and ultimate strength

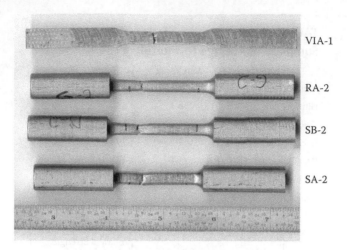

FIGURE 9.8 Example tensile specimens after testing.

(36 samples, across all member types) measured from Penns Creek Bridge are both significantly less than the assumed AASHTO values. The coefficients of variation (COV) from the test data are 22% for yield and 16% for ultimate, whereas the AASHTO procedure is based on a COV of only 10% for steel members. Both the reduced mean and increased COV would reduce the calculated reliability, or load rating, of a bridge. Since only three tests were performed on each member type, a meaningful COV cannot be calculated for each member type. Based on the data shown in Figure 9.7, one would expect each member type alone to exhibit a significantly smaller COV than that for all of the samples combined. The Penns Creek data reveals that a single wrought iron bridge may have widely variable tensile properties among its members. Although the AASHTO load rating procedure does not directly address this issue, it may be important to consider such variability in evaluating the strength of wrought iron bridges, especially in the case of non-redundant members or connections.

The results summarized in Table 9.2 and Figure 9.7 have some significantly lower values for both yield and ultimate strength compared to the data cited above. The size effect discussed by Withey and Aston[20] in which additional hot working required for members with smaller characteristic thicknesses increases yield and ultimate strength is somewhat evident in the test data shown in Figure 9.7. In general the large dimension square bars and round rods tend to have lower yield and ultimate strengths as compared to the flat plates, I- and C-sections tend to have greater yield and ultimate strengths.

Figure 9.8 shows examples of fractured tensile specimens from three different members: VIA-1, RA-2, SB-2, SA-2. The specimens from the 1-1/4" square bars (SA) exhibited low yield strengths and particularly low ultimate strengths, with very little ductility. Visual inspection of the specimens after testing did not reveal any inherent flaws or large slag inclusions. Figure 9.8 also shows that the fracture surfaces for specimen SA-2 are not parallel. For this, and other specimens, the fracture was observed to begin on one face and progress gradually across the width of specimen,

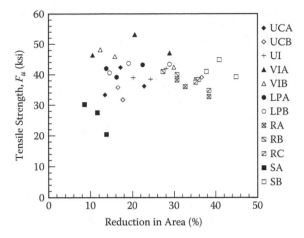

FIGURE 9.9   Tensile strength vs. reduction in area.

in contrast to the behavior of modern steels for which the fracture occurs essentially instantaneously. The gradual progression of fracture also resulted in some bending of the specimen.

Percent elongation and percent reduction in area were measured for each test; the results are summarized in Table 9.2. Wrought iron typically shows a wide variation in ductility as measured by percent reduction in area or elongation. The reduction in area varied from 11 to 41% with an average of 25%. The elongation (measured over a 1 in gage length) varied from 7 to 29% with an average of 20%. Figure 9.9 plots percent reduction in area vs. tensile strength for individual tests. Gordon[11] and Gordon and Knopf[13,14] report reduction in areas in the range of 20% to 60% and elongation from 11% to 30% for both historical and modern tests on samples from numerous sources. Bright[7] reported a mean elongation of 15%. The wrought iron from Penns Creek Bridge exhibited elongations and reductions in area consistent with those reported by other others.

HARDNESS TESTS

The average results of the hardness tests are summarized in Table 9.1; individual test results are shown in Figure 9.10, including both yield and tensile strengths. Average Rockwell B hardness ranged from 55 to 74 across all specimens with a mean of 65; Brinell hardness (BHN) ranged from 111 to 147 with a mean of 127. Sparks[18] cites a typical range of 95 to 130 BHN for wrought iron, with local values as high as 160. Elban et al.[10] report Rockwell B hardnesses in the range of 71 to 84 from a wrought iron I-section. The hardness results from Penns Creek Bridge are generally consistent with this data.

Elban et al.[10] cites a complicated procedure to estimate tensile strength from Brinell and Rockwell B hardnesses. Gordon and Knopf[14] states that hardness is often well correlated with strength in homogenous metals such as steel, but that hardness does not correlate well with strength or ductility of wrought iron due to the inhomogeneous nature of the fibers and slag. For low-alloy structural steels Sparks[18]

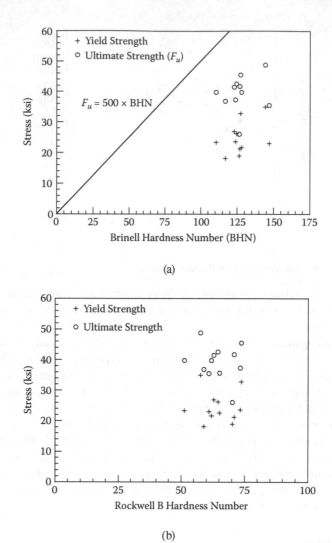

FIGURE 9.10 Yield and tensile strengths vs. hardness: (a) Brinell hardness, (b) Rockwell B hardness.

provides the relationship $F_u = 500 \times \text{BHN}$ where the result is in psi. This correlation allows field hardness measurements to serve as a convenient non-destructive test method for estimating tensile strength of steel. Figure 9.10 shows plots of yield and ultimate strength vs. Brinell or Rockwell B hardness. The plots show significant scatter and confirm the lack of correlation between strength and hardness in wrought iron. Even the Brinell test, which uses a larger diameter indentation and therefore might be less influenced by local variations in the material structure, exhibits no significant correlation with yield or ultimate strength.

Gordon and Knopf[14] also states that standard conversions between the various hardness scales (ASTM E140) do not hold for wrought iron. The hardness data

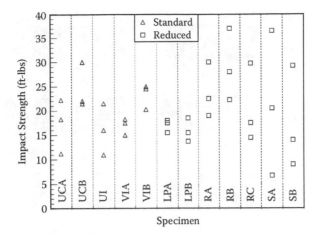

FIGURE 9.11   Impact strengths of individual specimens.

collected from the wrought iron of Penns Creek Bridge also confirmed that there was poor correlation between Rockwell B and Brinell hardness measurements from the same specimens.

Impact Tests

Table 9.1 summarizes the impact strengths measured with the Charpy V notch specimens. The average impact strengths ranged from 16 ft-lbs to 29 ft-lbs with a mean of 20 ft-lbs. Figure 9.11 shows the individual test results for impact strength. The specimen size (standard vs. reduced) did not significantly affect the range or average impact strength. Nearly all of the impact strengths measured from the wrought iron of Penns Creek Bridge were below the range of 25 ft-lb to 105 ft-lb reported by Sparks and Badoux.[19] For grade 36 steel in Zone 2 (minimum service temperatures from −1 to −30 degrees F) AASHTO[2] requires a fracture toughness of 25 ft-lbs (for fracture critical members) or 15 ft-lbs (for non-fracture critical members). Many of the specimens from this test clearly would not have met this criteria. According to Gordon and Knopf[14] and Sparks[18] wrought iron is typically considered to not be fracture critical, due to the ability of the fibers to arrest cracks. However, very few impact test results on historic wrought iron have been reported in the literature.

Some of the lowest individual impact strength test results were from samples from the square bars—7 ft-lb (specimen SA-3) and 9 ft-lb (specimen SB-2). Figure 9.12 shows the fracture surface of the SA and SB specimens. The fracture surfaces are characterized by a combination of a dark gray fibrous area and a shiny metallic faceted area. Gordon[11] discusses the appearance of the fracture surface as a traditional method of judging wrought iron quality. The reflective parts of the fracture surface are iron grains embrittled by the presence of phosphorous, and the source of the 19th century term of "crystallized" iron. The gray fibrous areas are locations where the individual fibers underwent a ductile fracture after significant elongation. This surface is similar in appearance to the failure surface from a tension specimen of wrought iron. The brittle fracture area generally starts at the notch. In the weakest

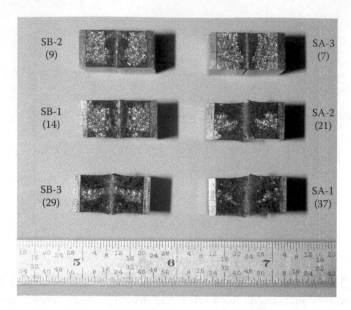

FIGURE 9.12   Impact fracture surfaces (value in parentheses is impact strength in ft-lbs).

specimens this brittle fracture covers a large percent of the cross-section. In the stronger specimens the brittle fracture area is bounded by the ductile fracture area in which the fibers have arrested the spread of the brittle fracture. In other specimens there was no apparent or visible brittle fracture area.

CONCLUSIONS AND FUTURE WORK

Wrought iron is known to exhibit great variability in its mechanical properties due to its inhomogeneity and manufacturing methods. Effective repair and rehabilitation of wrought iron structures requires specific knowledge of the behavior of wrought iron, which is substantially different than steel. The research presented in this paper aimed to provide additional data and further understanding of the mechanical properties of wrought iron and their variability, based on specimens salvaged from Penns Creek Bridge. Penns Creek Bridge was deigned and constructed in 1886 by the Columbia Bridge Works and the wrought iron was produced by Jones and Laughlins, Ltd. Tension, impact and hardness tests were conducted on variety of specimens—large dimension round and square bars, flat plates and I and C rolled sections.

Of particular importance to engineers evaluating the strength of existing wrought iron bridges is the observation that common mechanical properties may exhibit significant variation between members. In many samples the yield and ultimate strengths from the tension tests fell below values cited by other researchers, both historical and modern. Some of the lower tension strengths were associated with the large square bars, possibly due to the greater inhomogeneity of the slag as a result of less hot working during production. The specimens exhibited a wide variation in ductility, measured by elongation or reduction in area, as observed by previous researchers.

The Rockwell B and Brinell hardness measurements showed no significant correlation with yield or ultimate strengths. The approximate relation for hardness and ultimate strength used for early steels is not appropriate for wrought iron and is an under-conservative predictor of strength. The standard conversions between hardness scales are not appropriate for wrought iron. The measured Charpy V notch impact strengths of wrought iron varied widely. The fracture surfaces were characterized by an area of brittle fracture and an area of ductile fracture, and the lower strength specimens had a large relative area of brittle fracture.

For steel, field hardness testing can often be used as a simple non-destructive method to estimate strength. For wrought iron, a different method of field non-destructive testing remains an important research need. One potential approach would be to use a hand-held, x-ray fluoresence, metal analyzer which can determine the percentages of various chemical components.

The behavior of wrought iron depends heavily on its chemical composition and microstructure. In the future we hope to perform chemical analysis of the specimens to determine the quantities of components such as phosphorous, sulfur and carbon, which have been previously related to mechanical properties. We also hope to analyze the microstructure of the specimens and failure surfaces under a scanning electron microscope.

The preservation and rehabilitation of remaining wrought iron structures in the United States depends on continued research into the chemical, material and mechanical properties of wrought iron. Wrought iron structures represent an important period in the development of engineering in the United States and their historical significance goes beyond their function as buildings or bridges.

ACKNOWLEDGMENTS

The authors gratefully acknowledge the financial support provided by the Michael Baker Corp. through a summer undergraduate research fellowship at Bucknell University. The authors acknowledge the assistance of the Pennsylvania Department of Transportation, in particular Matt Hamel and Kara Russell. Jon Guizar of Nestlerode Contracting Co. Inc. arranged for salvage and transportation of the samples. Jim Gutelius and Tim Baker of Bucknell University assisted with testing and machining specimens.

REFERENCES

1. American Association of State Highway and Transportation Officials (AASHTO), "Manual for Condition Evaluation and Load and Resistance Factor Rating (LRFR) of Highway Bridges," 2003.
2. American Association of State Highway and Transportation Officials (AASHTO), "LRFD Bridge Design Specifications," 4th edition, 2007.
3. ASTM International, "Test Methods for Tension Testing of Metallic Materials" Standard E8, 2003.
4. ASTM International, "Test Methods for Brinell Hardness of Metallic Materials" Standard E10, 2003.
5. ASTM International, "Test Methods for Rockwell Hardness and Superficial Rockwell Hardness of Metallic Materials" Standard E18, 2003.

6. ASTM International, "Test Methods for Notched Bar Impact Testing of Metallic Materials" Standard E23, 2003.
7. Bright, S. "Strength of Victorian Wrought Iron Rail Bridges" message posted to Civil Engineering History & Heritage Exchange, <http://knowledgelists.ice.org.uk/ archives/ civil-engineering-heritage-l.html> Sept. 18, 2007.
8. Cultural Heritage Research Services, Inc. "T-510 Bridge over Penns Creek." Information Package prepared for U.S. Dept. of Transportation, Federal Highway Administration, and Pennsylvania Dept. of Transportation, January, 2005.
9. DeLony, E. and Klein, T. H. "Rehabilitation of Historic Bridges." *Journal of Professional Issues in Engineering Education and Practice*, 131 (3), 2005, 178–186.
10. Elban, W.L.; Borst, M.A.; Roubachewsky, N.M.; Kemp, E.L. and Tice, P.C. "Metallurgical Assessment of Historic Wrought Iron: U.S. Custom House, Wheeling, West Virginia." *APT Bulletin*, 29 (1), 1998, 27–34.
11. Gordon, R.B. "Strength and Structure of Wrought Iron." *Archeomaterials*, 2 (2), 1988, 109–137.
12. Gordon, R.B. *American Iron 1607–1900*, Johns Hopkins Univ. Press, Baltimore, Maryland, 1996.
13. Gordon, R. and Knopf, R. "The Aldrich Change Bridge: Evaluation of the Strength of Historic Bridge Iron." *IA: J. of the Society for Industrial Archeology*, 28 (2), 2002, 5–14.
14. Gordon, R. and Knopf, R. "Evaluation of Wrought Iron for Continued Service in Historic Bridges." *J. of Materials in Civil Engineering*, 17 (4), 2005, 393–399.
15. Jardini, D. "From Iron to Steel: The Recasting of the Jones and Laughlins Workforce between 1885 and 1896." *Technology and Culture*, 36 (2), 1995, 271–301.
16. Kemp, E.L. "The Introduction of Cast and Wrought Iron in Bridge Building." *IA: J. of the Society for Industrial Archeology*, 19 (2), 1993, 5–16.
17. Pennsylvania Department of Transportation (PennDOT) "TR 510 over Penns Creek" BMS# 14721705100005. *Pennsylvania Historic Bridge Inventory and Evaluation*. Prepared by A. G. Lichtenstein and Associates, Inc., 2003.
18. Sparks, P. "Guide to Evaluating Historic Iron & Steel Bridges" Report prepared for the Texas Department of Transportation, 2004.
19. Sparks, S. P. and Badoux, M.E. "Non-Destructive Evaluation of a Historic Wrought-Iron Truss Bridge in New Braunfels, Texas." *APT Bulletin*, 29 (1), 1998, 5–10.
20. Withey, M.O. and Aston, J. *Johnson's Materials of Construction*, 6th ed. John Wiley & Sons, Inc., New York, 1926.

Part 4

Preservation, Rehabilitation and Restoration

10 The Preservation of Historic Bridges

Allan King Sloan

CONTENTS

Thanks to enthusiastic supporters of historic bridge preservation, there is an emerging awareness of the value of saving old bridges on the part of citizens and communities all over the country. While not designed and built to serve the needs of modern traffic, many of these structures continue to provide useful service and are recognized as amenities rather than problems or hazards to the communities in which they are located. Many of these old bridges may be an integral part of local history or may be historic in terms of their design and engineering characteristics, or both. Efforts to preserve them has not been easy and, indeed, most of the iron and steel bridges in the country built in the great period of economic growth and expansion after the Civil War have long since disappeared. However, there are some interesting examples of successful efforts at preservation which are worthy of mention and possible emulation.

This paper will focus on the efforts to preserve bridges built by the King Bridge Company of Cleveland, Ohio. It was founded by my great-great grandfather, Zenas King, in 1858, and later run by my great grandfather, James A. King and his younger brother, Harry Wheelock King, after Zenas died in 1892. By the late 1890s the company claimed to have built over 10,000 bridges all over North America. The company started building simple wrought-iron bowstring bridges in the 1860s and 70s, then went on to build standard trusses and a variety of movable bridges later on. They became a specialist in cantilever bridges and even built a notable suspension bridge in St. Louis. By the turn of the century, they had built a number of bascule bridges, trestles, and scores of solid beam girders for the railroads. During the six decades the company was in business, it experienced vast changes in the technology and business operations of the independent bridge builders and by the early 1920s, its functions had been taken over by the big steel companies, highway departments, and large civil engineering firms, some of which were "spun out" of the company.

However, King bridges from each of these eras still exist. Over the past few years a number have been preserved. Some have been fixed up to continue to serve vehicular traffic; some have had traffic removed and are now used as key elements of parks and other local amenities; some have been physically removed to a new location to provide a new and different function; some bridges built for the railroads now serve as features of hiking and biking trails; and some historic bridges have been carefully maintained to serve traffic as well as to represent the engineering and design concepts of earlier bridge-building art.

In each of these cases, a particular dynamic of socio-political forces have come into play that have made the project successful. While the technical and engineering solutions to preserving old bridges are often relatively easy and straightforward, the "politics" of preservation is often more complicated. Yet it is the essential ingredient in a successful preservation program. By far the easiest course for the owner of most old bridges be it a town, county or state highway department (or a railroad company) is to remove them when no longer able to carry modern traffic. The tougher decision is to preserve them which may involve safety and liability issues as well as pressure to "modernize". The following are the stories of some old bridges that have successfully dodged extinction and have found new life.

SCENARIO #1: FIX-UP INSTEAD OF REPLACEMENT

There are some communities in which the local highway authorities, with the strong support of the affected community, decided to fix up old bridges to continue to carry normal traffic instead of replacing them. While this is generally a rare occurrence, there are three recent examples that buck the trend.

In Hopewell Township, New Jersey, the oldest bridges in this suburban community in Mercer County (near Princeton and Trenton) are two King through trusses which "history-conscious" local citizens wanted to keep operating instead of replacing with modern structures. The Bear Tavern Road Bridge (Figure 10.1) built in 1882 is a Pratt truss that carries a relatively high volume of auto traffic for an old bridge but has remained in good enough shape for the Mercer County Highway Department to reinforce abutments and replace the stringers and flooring to keep it in operation. The Mine Road Bridge built in 1885 is the other Pratt truss that still carries vehicular traffic and has needed little structural alteration over the years. Both bridges have earned the affection of the local citizenry and have been the subject of study by local school children interested in their preservation. Given their status in the community, it was easy for the Mercer County highway engineers to justify their rehabilitation instead of replacement as the best solution.

In Marion, Virginia, an 85-foot Pratt through truss called Happy's Bridge (Figure 10.2) built by the King Bridge Company in 1885 connects a road intersecting Main Street in downtown across a small river with a riverside park and some other public buildings. The local community decided it wanted to keep rather than replace the old bridge and in 2005 it was rehabilitated as a joint venture of Virginia

(a)

(b)

FIGURE 10.1 The Bear Tavern Road Bridge during and after rehabilitation. (*Source: Charlotte Pashley and A.K. Sloan.*)

DOT and the Town of Marion for total project costs of $481,088, 80% funded by the state through Federal T-21 grant money and 20% by the town. The reopening of the bridge was a cause for a community celebration that provided an excuse for owners of old wagons and cars to proudly parade their vehicles.

In Lewis and Clark County, Montana, the Dearborn River High Bridge (Figure 10.3) is a unique four span 160 foot deck truss listed on the National Register of Historic places. It was built by the King Bridge Company in 1897. Located in a spectacular site on a remote county road in the foothills of the Rockies, it is not particularly well known to local inhabitants who would have little reason to use it. However, the historian of the Montana Department of Transportation, Jon Axline,

(a)

(b)

FIGURE 10.2   Happy's Bridge before and after restoration. (*Source:* A.K. Sloan and S. Wilson —Thompson&Litton.)

FIGURE 10.3   The Dearborn River deck truss restored. (*Source:* Montana State Department of Transportation.)

was able to persuade his peers that this historic structure was well worth preserving, and in 2003 the department contracted with HDR Engineering to do the repair of the truss components, piers, decking and abutments with spectacular results.

The more likely scenario to preserving an old bridge is by insuring that the structure is taken out of harm's way; that is removing it from an active of vehicular traffic system This can occur either if the location of the bridge lends itself for a non-traffic use as part of an "amenity" or if it can be relocated to a "safe" location. There are a number of recent examples of each of these situations.

(a)

(b)

FIGURE 10.4   The Stuart Road Bowstring restored. (*Source:* A.K. Sloan and J. Stewart.)

SCENARIO #2: USING THE BRIDGE IN AN "AMENITY PACKAGE"

A number of communities have used their old bridges in projects to create an "amenity package." This may include reducing or eliminating their use for vehicular traffic, turning them into pedestrian facilities, or integrating them into parks or public spaces. Three of these have been undertaken recently in upstate New York.

In Chili Mills, Monroe County, the Stuart Road Bridge (Figure 10.4) is 74-foot bowstring sitting adjacent to a picturesque mill pond surrounded by the original buildings carefully tended to for years by the Wilcox family, the owners of the mill site. It has been known to the locals as the "Squire Whipple" bridge in honor of the inventor of this bowstring design and was built by the King Bridge Company in 1877. It has played a role in an annual village celebration of "the Squire" in full period

(a)

(b)

FIGURE 10.5   The Beach Road Bowstring before and after restoration. (*Source:* A.K. Sloan and N. Holth.)

dress. After persistent efforts of the Wilcox family and their friends, the Monroe County Department of Transportation undertook the rehabilitation of the bridge in 2002 using their own manpower and at minimal cost.

In Newfield, Tompkins County, the Beech Road Bridge (Figure 10.5) is a 54-foot Zenas King patented Bowstring and one of two historic bridges in this village near Ithaca. The covered bridge in the center of the village has long been celebrated. The bowstring built in 1873 had not been subject to the same esteem until recently. Long closed to vehicular traffic, the bowstring plays and important role as a pedestrian crossing of a deep ravine, particularly for school children. For years the responsibly for the upkeep of the bridge was debated between village and county officials until a local ad hoc citizens group headed by local citizen, Karen Van Etten, organized the effort to

(a)

(b)

FIGURE 10.6   The Grasse River Bowstring rehabilitation underway. (*Source:* J. Stewart and Grasse River Heritage.)

rehabilitate the structure. After years of lobbying and fund raising, the bowstring was and rehabilitated in 2004 and the grand reopening covered by the *New York Times*. The project costs of $77,000 were contributed by the county with funding from Historic Ithaca, the local historic preservation organization. As a follow up, a local land owner has contributed property next to the bridge for a new village park.

In Canton, St. Lawrence County, the Grasse River Bowstring (Figure 10.6) is a long-abandoned 1870-vintage King tubular arch bowstring across the Grasse River near the center of this college town. It has the potential to provide pedestrian access to an island in the middle of the river which a local preservation group, the Grasse River Heritage Area Development Corporation, is creating a park celebrating the industrial history of the town. Rehabilitation will allow the bridge to be used as pedestrian access to river islands once populated by mills. The bridge and island restoration was funded by a grant of $177.353 from New York State and $110,000 raised by the Development Corporation from local citizens. Barton and Loguidice

(a)

(b)

FIGURE 10.7 The New Bridge at River Edge before and after restoration. (*Source:* Bergen County Historical Society and A.K. Sloan.)

Engineering of Syracuse managed the project starting with the rehabilitation of the bowstring completed in November, 2007.

In River Edge, Bergen County, New Jersey, a 110-foot Pratt swing bridge was built by the King Bridge Company 1889 and known ironically as "New Bridge" (Figure 10.7). It is listed on the National Register of Historic Places. Owned by the county, it was rehabilitated some years ago to serve as a pedestrian crossing over the Hackensack River. It connects to the headquarters of the Bergen County Historical Society at Steuben House, an important historical site dating from the War for Independence to a town park on the other side of the river. Although now stationary, the mechanical elements of the turntable used to move the bridge (by hand) are still in place.

(a)

(b)

FIGURE 10.8   The old spans at Wellsbridge beside the replacement bridge. (*Source:* J. Stewart and A.K. Sloan.)

In Unadilla, New York, a two-span Pratt through truss bridge crosses the Susque-hanna River in the hamlet of Wellsbridge (Figure 10.8), located adjacent to State Route 44. When the bridge became unsuitable for modern traffic, state highway officials, in their wisdom, decided to leave the old bridge in place and build the new bridge in a parallel alignment. The result is an interesting combination of old and new structures with the King-built 1886 trusses providing a pedestrian crossing and viewpoint for river watching. For "safety" reasons, the old bridge's capacity has been restricted to only eight people at a time. In the view of some local bridge enthusiasts, this prevents its use as a place to watch rafting and other river sports, a major attrac-tion in the area.

SCENARIO #3: MOVING TO A NEW SITE

A number of interesting examples of old bridges were moved to new locations to insure their preservation. These efforts pose a number of logistical problems and sometimes require heroic efforts, but the results are often spectacular.

In Jones County, Iowa, the Hale Bridge (Figure 10.9) a Zenas King patent bowstring comprising two 80-foot spans and one 100-foot span was built in 1879 and listed on the National Register and in HAER. On Wednesday, March 8, 2006, Iowa Army National Guard Chinook helicopters moved the rehabilitated trusses from the staging site to their new home at the Wapsipinnicon State Park in Anamosa This landmark event drew an excited crowd of Iowans and was covered by the History Channel's new series MEGA MOVERS that was aired on June 27, 2006, as well as the *New York Times* and the local press. The restoration was completed in late summer of 2006 and the bridge now serves as the new entrance to a hiking and biking trail in the park. The Jones County Historical Society headed by Rose Rohr took the lead in organizing and orchestrating this highly successful multi-year bridge preservation effort in which a large number of state and local governmental agencies were involved.

Ashtabula County in northeastern Ohio is known for its historic covered and iron bridges. The Mill Creek Road Bridge (Figure 10.10) is a 104-foot Pratt Through Truss built by the King Bridge Company in 1897 that was rehabilitated and relocated from Mill Creek Road to the Western Reserve Greenway Trail. The project was supervised by the Ashtabula County engineer's office and Union Industrial Co. of Ashtabula was the contractor. The project cost included $81,311 for removal and disassembly and $ 209,570 for structural rehabilitation for a total of $291,000. The Grand River Partnership, a private group devoted to the protection and enhancement of the rivers in northeastern Ohio, hopes to protect a similar bridge on Johnson Road by including it in a scenic easement being acquired on adjacent land along the river.

In Northport, Alabama, the Black Warrior (Espy) Bridge (Figure 10.11) is a single 203-foot bowstring built in 1882 as part of a three span bridge across the river that was removed many years ago to a remote location in the county. It is the oldest iron bridge in the state. Now it is being relocated back to near its original location on the levee system in Northport as part of a walking trail system. Funds are being provided by the Alabama DOT using Federal T-21 money with 20% to be provided by the City of Northport. This effort has required years of hard work by the Friends of Historic Northport, a local citizens group led by Ken Willis and others which had to develop the concept for relocating the structure, raise money, and persuade the public authorities to undertake the project. Plans for the disassembly and moving of the bridge have been completed and the project is finally underway at an estimated cost of about $115,000. The civil engineering department of the University of Alabama is also assisting in the program.

While moving an old bridge to a new location requires a substantial logistical effort, local acceptance, and cost, there are other situations in which an old structure abandoned for its original use can still serve a new function. This is particularly true of railroad bridges that have been left standing after the rail services have been terminated and current owners are willing to change the function of the structure.

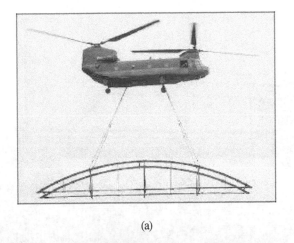

(a)

(b)

(c)

FIGURE 10.9   The spectacular relocation of the Hale Bridge, Anamosa, Iowa. (*Source:* A.K. Sloan, Cedar Rapids Gazette, and Jones County Historical Commission.)

(a)

(b)

FIGURE 10.10   The Relocated Mill Creek Bridge, Ashtabula County. (*Source:* A.K. Sloan.)

SCENARIO #4: CREATIVE USE OF OLD RAILROAD STRUCTURES

As the importance of nation's railroads have faded, there are a number of abandoned or underutilized bridges of various types that have been put to new use, particularly as part of "rails to trails" and similar programs. A number of old King bridges have been in this situation.

In Venango County, Pennsylvania, the Belmar Bridge (Figure 10.12) is a 1,361-foot long structure built for the Jamestown, Franklin and Clearfield Railroad in 1906 by

FIGURE 10.11   The Black Warrior Bridge awaiting relocation. (*Source:* K. Willis.)

FIGURE 10.12   The Belmar Bridge across the Allegheny River. (*Source:* D. Alward—Venangoil website.)

the King Bridge Company under a subcontract to the Thomas McNally Company of Pittsburgh. The bridge is now part of the East Sandy Creek Bicycle Trail operated by the Allegheny Valley Trails Association and offers panoramic views of the Allegheny River and the surroundings.

There are other examples of rail to trail conversions, including the Tunnel Hill State Trail in Southern Illinois between Harrisburg and Karnak. Five King bridges originally built in 1912 for the Old Big Four Railroad are now used by hikers and bikers through one of the most picturesque areas of the state. The trail itself recognizes a variety of railroad bridge engineering.

In Ulster County, New York, the King Bridge Company built a 925-foot trestle across Esopus Creek in Rosendale for the Wallkill Valley Railroad in 1895, called the Rosendale Viaduct (Figure 10.13). When all service on this line was abandoned in 1976, a local railroad enthusiast and entrepreneur, John Rahl, used his research of state law

(a)

(b)

FIGURE 10.13   The Rosendale Viaduct pictured in an 1890s King Bridge Co. catalogue and today. (*Source:* A.K. Sloan.)

regarding railroad abandonment to purchase the structure and eleven miles of adjacent rail bed for a minimal cost and converted the viaduct for pedestrian use as part of the Wallkill Valley Rail Trail. The bridge is in good condition and has been equipped with wooden planking and railings so that people can walk out onto the bridge to take in the spectacular views across the valley and the Hudson River beyond. The viaduct is considered to be unique landmark and an asset to the village of Rosendale.

In St. Francisville, Illinois, the Wabash Cannonball Bridge (Figure 10.14) built in 1906 once carried the famous Wabash Cannon Ball train across the Wabash River. When the railroad abandoned the line, the bridge was purchased by a local farmer to haul his produce across the river but is now owned and maintained by the town of St. Francisville as an historic artifact. Its one lane is still open for trans-river vehicular traffic.

Two imposing bridges built for the New York Central Railroad on the Buffalo to Rochester line by the King Bridge Company are still standing. The first is a 124-foot

FIGURE 10.14   The Wabash Cannonball Bridge today. (P. Kennedy.)

FIGURE 10.15   The Deck Truss at Lockport. (*Source:* A.K. Sloan.)

Deck Truss Bridge (Figure 10.15) across the New York State Barge (Erie) Canal in Lockport, near Buffalo. It was built in 1902 and is still used today for occasional passenger excursion and local freight trains. Tom Callahan owns the old water works facilities adjacent to the canal for development as an exhibition of historic hydraulic technology. He is leading efforts to restore the footbridge along side the track, traditionally one of the best places to view the five step locks of the canal, one of the areas important tourist attractions. The second is the 304-foot Hojack Swing Bridge (Figure 10.16) near the mouth of the Genesee River in Rochester, built in 1905 and abandoned in 1993. Despite the valiant multi-year effort of a group of local preservationists headed by Richard Margolis, the U.S. Coast Guard has ordered the removal of the bridge, as it is no longer used for transportation and is a "hindrance to navigation". Public officials in Rochester have shown remarkable indifference to the preservation of this fine example of swing bridge technology. Since the costs to the

FIGURE 10.16   The Hojack Swing Bridge at Rochester. (*Source:* A.K. Sloan.)

owner (CONRAIL) of its removal will be substantial and the impact on the river of the removal of the turntable substructure unknown, the bridge is still in place.

Fortunately, there are communities that appreciate the value of their old bridges and have taken measures to insure their protection and continued use, even if extensive maintenance and rebuilding is required. Cleveland, the home of the King Bridge Company, and New York City, which was in effect created and developed by its historic bridges. Two King built bridges represent the best of this tradition.

SCENARIO #5: CARE AND MAINTENANCE OF IMPORTANT BRIDGES

In Cleveland, Ohio, the Center Street Swing Bridge (Figure 10.17) is a famous bob-tailed swing bridge built in 1901. It is now part of the Cleveland's impressive inventory of historic bridges, three of which were built by the King Bridge Company. It still functions as a vehicular crossing of the Cuyahoga River providing access to the entertainment complex in the Cuyahoga River Flats. It is historically important, both for its design and role as a working swing bridge. It is often described in historic bridge literature and is kept in operation through the enlightened maintenance program of the city's bridge engineering department.

In New York City, the University Heights Swing Bridge (Figure 10.18) crossing the Harlem River at 207th Street in Manhattan to West Fordham Road in the Bronx, began life as a swing bridge across the Harlem Ship Canal at Knightsbridge Road in 1895 and was featured in the King Bridge Company catalogues of that era. To make room for a larger bridge that would carry the Broadway subway line across the canal, this bridge was loaded on barges and floated to its present site in 1905 and reconstructed with new piers and approaches. The bridge was considered to be a significant engineering and architectural structure was awarded landmark status in 1983 by the City's Landmark Preservation Commission. The bridge now serves both vehicular and pedestrian traffic moving between the Inwood Community in Upper Manhattan and the Fordham University area in the Bronx, and a visit to the bridge is well worth the experience.

FIGURE 10.17   Cleveland's Center Street Swing Bridge today. (*Source:* W. Vermes.)

SUCCESS FACTORS IN OLD BRIDGE PRESERVATION

These examples of preservation efforts each represent one or a variety of factors that have been key to a successful outcome. While these examples are selective and do not necessarily represent the universe of successful programs, they do demonstrate the main characteristics that appear to be essential. These are as follows:

1. The Need for a "Champion." It is clear that these preservation efforts would never have been mounted unless there was a "champion" leading the charge. This champion might be a local historical or environmental group, a dedicated individual with lots of energy, patience and fortitude, an enlightened local highway department, state DOT, or an engineering firm willing to undertake projects with often modest funding. These champions must be willing to work hard to understand the "politics" of the situation, organize community support, find funding sources, pull strings, and make sure the process works. Without such a champion, most efforts will fail.

2. An Appropriate "Setting" and "Environment." There needs to be the realistic opportunity for the old bridge to be put out of harms way (i.e., not put in the position of continually having to carry high volumes of modern traffic, unless extensively rebuilt). Public parks or reservations, riverside conservation areas, hiking-biking trails give the old bridge a chance to become a local "amenity" rather than a traffic bottleneck, a danger point, or an eyesore. If the bridge is still to be used for traffic, restraints and restrictions have to be honored, particularly by the drivers of large and heavy vehicles.

3. A "Sympathetic and Supportive" Local Community. Local village, town, city or county officials must be supportive rather than hostile to the preservation effort. They must recognize the potential role the bridge can play in enhancing the community's image and in celebrating its history The indifference or hostility of local, county or state highway officials can be a serious impediment to old bridge preservation It is often the role of the "champion" to lobby for this support with whatever methods of persuasion are available.

FIGURE 10.18.1 The Harlem Ship Canal Swing Bridge pictured in the King Bridge Company Catalogue.

FIGURE 10.18.2 The University Heights Bridge today. (*Source:* A.K. Sloan.)

4. Funds for Preservation. In the preservation efforts noted above, the level of funding required for preservation have ranged from minimal (Chili Mills) to moderate (Mercer County's annual maintenance program for the Hopewell bridges) to major (for the large relocation efforts (in Iowa). Funds have come from a variety of sources with Federal Transportation Act T-21 funds playing a key role in large projects. Local taxes and funds raised by private non-profit organizations like historical and environmental groups are very important catalysts for obtaining public funding. Bundling old bridge preservation funding in with larger park development programs (Canton, New York) is often a good way to get adequate funding levels. However, unless those officials who control major funding sources are brought on board, the preservation efforts will be hard to achieve.

5. The Historic Bridge Fraternity Has an Important Role to Play. Putting the old bridge into its appropriate historical context is most often a key factor in justifying efforts to preserve it. The members of the historic bridge fraternity are the holders and purveyors of this information. Federally mandated state-wide historic bridge inventories are often a good starting point for a particular preservation program, but being included on an historic bridge list does not guarantee survival. The growing number of bridge preservationists are now connected through the internet. Websites created by Nathan Holth, James Baughn, Daniel Alward and others are extremely important in flashing the warning signals when an old bridge is in danger. Local champions need to be supported by the historic bridge fraternity in their efforts to justify preservation by providing "bridge history" to supplement to role of the bridge in "local history". Both types of justification are usually needed for success.

In the successful examples cited above all five of these factors were favorable. Where one or more of these factors are missing, the efforts are most often frustrated.

AUTHOR'S NOTE

In preparing this paper, the preservation examples selected are those in which I have some direct personal knowledge, including visits. Thus they are heavily concentrated in New York, Ohio, and other eastern states which are near my home base. There are many other examples that should be included in any complete list of successful preservation efforts, particularly in Texas, (The Faust Street, Moore's Crossing, Alton in Denton, and the Bullman Bowstring bridges) Indiana (The Boner Bowstring, Madison, and Atterbury bridges), Michigan (the bridges in Allegan and Belle Isle), Kentucky (the Singing Bridge in Frankfurt and the Bowling Green Bowstring), Minnesota, (the Merriam Street Bridge), Wyoming, (the Fort Laramie Army Bridge), as well as Iowa, Missouri, Nebraska, Kansas, Arkansas, and even Nova Scotia. Historic bridge preservation programs in many of the Midwestern states have very strong backing from State DOTs. This has been important in the number of successful programs implemented there.

11 Preservation of Historic Iron Bridges

Adaptive Use Bridge Project, University of Massachusetts–Amherst

Alan J. Lutenegger

CONTENTS

ABSTRACT

A project is underway at the University of Massachusetts–Amherst to refurbish and rebuild eight late 19th Century and early 20th Century iron highway bridges for use as pedestrian bridges on the University campus. The project will be completed over the next 10 years and is largely being performed by Civil Engineering students with assistance from private sources and without any state funds. The eight bridges, which include both pony truss and through truss bridges that have been scheduled for this project are described. A description of the preservation process is also given using one of the bridges that has been completed as an example. The project illustrates how the process can be used in engineering education, engineering outreach, service learning and historical engineering heritage preservation.

INTRODUCTION

Iron and steel truss bridges are rapidly disappearing from the landscape, being replaced by modern steel and concrete structures that have wider lanes and higher load capacity. At the University of Massachusetts, a project is underway to save these old bridges for reuse so that they can have a new life and provide service once again. The goal of the project is to obtain, restore, rebuild, and preserve a number of historic bridges for reuse on the UMass Amherst campus. The bridges will be used for a combination of pedestrian/bike paths and jogging/walking trails, across the campus. The preservation and restoration of historic bridges is not new and has been successfully demonstrated in the past. However, these are largely single event projects that have occurred at various individual sites.

Old highway bridges provide excellent opportunities for adaptive use. As older steel and iron bridges are slowly being replaced by more modern structures, new uses can be found for these structures. Descriptions of the adaptive use of old highway bridges to modern pedestrian bridges have been given by Zuk and McKeel (1981) and Green et al. (1990). Typical examples of adaptive use include uses in walking or hiking trails and bike paths since the new loads are generally much lower than original design loads for highways. Structural stability is typically not an issue with reuse of a bridge, provided that deteriorated or broken members are repaired or replaced. In addition to preserving the historical values of old bridges, these structures are aesthetically pleasing and can provide valuable service.

The UMass project is being undertaken by students in the Department of Civil and Environmental Engineering at UMass and the Student Chapter of the American Society of Civil Engineers (ASCE) and the Associated General Contractors (AGC) and will ultimately provide a Living History outdoor museum of historic bridges on the campus that would provide a number of potential uses. The Project will develop a walking tour of the campus to each of the bridge sites. The Project will be performed without any direct state funding and will rely on donations of cash, in-kind services, equipment, and materials and through private donations. The restoration and reconstruction of the bridges will be performed by the ASCE students over the next several years.

The bridges will primarily by reconstructed by students in the Department of Civil and Environmental Engineering and reconstructed at selected locations on the campus. The primary objectives of the Project are to: 1) preserve examples of civil engineering structures that would otherwise be demolished for scrap; and 2) use the refurbished bridges as engineering teaching aids. There are a number of supplemental objectives as well, including: 1) provide engineering students with an opportunity to obtain hands on design and construction experience; 2) provide students with an understanding of the History and Heritage of Civil and Environmental Engineering; 3) provide valuable historic preservation of early iron/steel bridge structures; 3) enhance the campus; 4) provide a format for university engineering outreach to

TABLE 11.1
Summary of Bridges Scheduled for Refurbishment

Bridge No.	Year	Span Length (ft.)	Style	Origin
1	1906	42	Warren Pony Truss with Verticals	VDOT
2	ca. 1895	52	Warren Pony Truss Without Verticals	Private
3	1886	40	Ball Wrought Iron Pipe Truss	Private
4	1885	81	Berlin Iron Bridge Co. Lenticular Pony Truss	MHD
5	1894	40	Single Intersecting Warren Pony Truss	MHD
6	1884	103	Berlin Iron Bridge Co. Lenticular Through Truss	MHD
7	1880	77	Wrought Iron Bridge Co. Pratt Pony Truss	MHD
8	ca 1892	73	Berlin Iron Bridge Co. Curved Chord Pony Truss	VDOT
9	1895	93	Wrought Iron Pratt Through Truss	MHD

Note: VTRANS—Vermont Transportation Agency; MHD—Massachusetts Highway Department.

K-12 students to learn more about engineering; 5) provide permanent structures for use as teaching aids in the CEE undergraduate and graduate curricula. The Project was initiated by the author in 1997 and has been developing steadily. A total of nine bridges have been donated for the Project. Table 11.1 gives a summary of the bridges that have been donated and delivered to the University for the Project.

The Project also provides another opportunity for CEE Students to develop engineering, management, and teamwork skills. Each bridge will be carefully documented by measuring and cataloguing the existing members. Students will inventory and measure each bridge member and identify those that need to be replaced. New parts will be fabricated and used to replace badly corroded members. The Project will also preserve an important era of Civil Engineering History and Heritage and will engage students in hands on activities outside of the classroom.

RECONSTRUCTION VERSUS RESTORATION

The first step of the Project was taken in the fall of 2001, the CEE Department took delivery of a 40 ft. long steel Warren Truss Pony Bridge donated by the Vermont Transportation Agency (Bridge No. 1). The bridge was built around 1906 and was first used as a highway bridge in southern Vermont. Prior to delivery of the bridge to Amherst, the bridge had been in storage for about 8 years by VTRANS. The bridge was delivered in four sections and was completely rebuild by CEE students at UMass. The bridge was inventoried; AutoCad drawings were prepared; replacement parts were fabricated to rebuild the trusses; and a design was completed for reconstruction of the bridge at an appropriate location.

The bridges are being preserved and reconstructed using modern materials and construction practices so that the original configurations of the bridges are preserved. There is no attempt to "restore" the bridges in the sense of using original techniques, such as riveting, using techniques that might closely resemble original practice. Instead, all components that are replaced are being fabricated using A36 steel and rivets are being replaced with high strength bolts. All removed components are

TABLE 11.2
Summary of Preservation Stages and Tasks

Stage	Task	Summary
I – Inventory/ Documentation	1. Inventory Bridge	Each bridge will be inventoried by students. All bridge components will be marked for identification and will be measured
	2. Develop Detailed Documentation	AutoCad drawings of each bridge will be prepared. This will provide a permanent record of the bridge components and will include general drawings and detailed drawings of individual components and connections.
II – Refurbishment	3. Fabrication of New Components	All deteriorated components will be replaced with new components. All machining will be performed at the College of Engineering Support Services machine shop when possible. In special cases it may be necessary to fabricate some components at an outside shop, however this is not anticipated at this point. All new components will be primered prior to use.
	4. Structural Analysis	For each bridge, a new design will be performed for the anticipated use and location of the bridge. At this time all bridges are expected to bridges are be used as pedestrian bridges.
III – Reconstruction	5. Replacement of Components	Initial restoration/reconstruction will consist of removal of deteriorated components and replacement with newly fabricated components. During this phase of the work some cleaning of existing members will also be performed.
	6. Foundation Construction	Reconstruction will include design and construction of new foundations, refurbishment of trusses, placement of new bridge beams and stringers, and installation of new decking and railing.
	7. Beams, stringers, decking, guard rails	Installation of beams, stringers and decking. Fabrication and installation of handrail to meet code requirements.

marked and inventoried so that they can be used in the future for laboratory testing of hardness and tensile strength.

Any construction/reconstruction project requires a series of steps or phases for successful completion. Table 11.2 provides a summary of the various stages of work being performed on each bridge and identifies individual tasks within each stage. Each task within each stage provides an opportunity for students to gain experience in different aspects of a Civil construction project.

DESCRIPTION OF BRIDGES

BRIDGE NO. 1.: WARREN PONY TRUSS

The first bridge obtained was a 40 ft. span Warren Pony Truss bridge was donated and delivered to UMass by the Vermont Transportation Agency for this Project.

FIGURE 11.1 Students working to remove old handrail from Bridge No. 1.

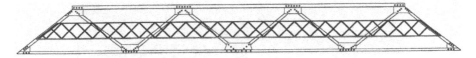

FIGURE 11.2 AutoCad Drawing of truss from Bridge No. 2.

The bridge was built in 1906 and was formerly located in southern Vermont. The bridge was removed in 1996 and was placed in storage by VTRANS for later use (see Figure 11.1). This bridge has been successfully rebuilt and is currently in use to span a small stream on campus on a walkway.

BRIDGE NO. 2: WARREN PONY TRUSS

An iron/steel Warren Pony Truss Bridge, complete with steel safety railing, was donated to the CEE Department by a private owner for use on this Project. The bridge was essentially complete except for the decking and is currently in outside storage. The bridge was formerly in service as a highway bridge in Northfield, Ma. and has a span of 52 ft. It is different from Bridge No. 1 in that it has no vertical members. The bridge is currently undergoing reconstruction and should be completed in the spring of 2008. (See Figures 11.2 and 11.3.)

BRIDGE NO. 3: BALL PIPE KING POST BRIDGE

It is estimated that as many as 40 Ball Pipe Truss Bridges were manufactured in E. Windsor, Ma. in the late 1800's by the local entrepreneur Charles Ball. Only three of these bridges are known to survive. The Ball Pipe King Post Bridge shown below was originally located on Holiday Road in Dalton, Ma. but was replaced with a new structure in 1990 (see Figure 11.4). The bridge was taken over by the Windsor Historical Commission and has been in open storage since then. This bridge is of the King Post design with a span of 40 ft. The Windsor Historical Commission

FIGURE 11.3   Trusses of Bridge No. 2 as originally found.

FIGURE 11.4   Ball pipe bridge as originally found.

transferred ownership of the bridge to the University for this Project. The bridge is being dismantled and moved in small sections for refurbishment.

Bridge No. 4: Lenticular Pony Truss Bridge

At one time, there were approximately 600 lenticular truss bridges built and erected throughout New England by the Berlin Iron Bridge Co. of East Berlin, Ct. (see Figure 11.5). Only about 60 of these bridges remain, of which only nine are located in Massachusetts. The Golden Hill Road bridge over the Housatonic River in Lee, Ma. is an example of one of the few surviving Lenticular Pony Truss Bridges in the country. The bridge has a span of 76 ft. The Massachusetts Highway Department (MHD) transferred ownership of this bridge to the University for this Project. The bridge was dismantled and transported to UMass by the contractor in 2005 under a Special Provisions clause in the bridge construction contract.

FIGURE 11.5   Berlin Iron Bridge Co. lenticular pony truss bridge.

FIGURE 11.6   Bridge No. 5 while still in service.

BRIDGE NO. 5: SINGLE INTERSECTION WARREN PONY TRUSS

The Reed's Bridge Rd. Bridge connecting Bardwell's Ferry Rd. and Reed's Bridge Rd. in Conway, Ma. was built by R.F. Hawkins in 1894 (see Figure 11.6). The bridge has been replaced by a new structure by the Massachusetts Highway Department. This bridge is a single intersecting Warren Pony truss bridge with a span of approximately 40 ft. MHD agreed to transfer ownership of this bridge for this Project. The bridge was dismantled and transported to UMass by the contractor in 2005 under a Special Provisions clause in the bridge construction contract. The bridge has a number of interesting structural features, including an open steel grate deck system.

BRIDGE NO. 6: LENTICULAR THROUGH TRUSS

Another example of a Lenticular Truss of the through truss design is the Galvin Road bridge spanning the Hoosic River in North Adams, Ma. (see Figure 11.7). This

FIGURE 11.7   Berlin Iron Bridge Co. through truss bridge.

bridge is unique in that it is only one of two surviving single span through truss
bridges of this type in Massachusetts. The other bridge of this type is the Bardwell's
Ferry Bridge located in Shelburne, Ma. and is listed on the National Registry of
Historic Places and on the American Society of Civil Engineers Historic Structures
List. The bridge has a span of 90 ft and was replaced by a new structure in 2005.
The Massachusetts Highway Department (MHD) agreed to transfer ownership of
this bridge for this Project. The bridge was completely dismantled and transported
to UMass by the contractor under a Special Provisions clause in the bridge construction contract. This bridge represents the largest span length of all candidate bridges
and the most complex and will require the largest amount of time for refurbishment
and reconstruction. It is expected that this bridge would become the showpiece of
the collection.

BRIDGE NO. 7: PIN-CONNECTED PRATT PONY TRUSS

The Lower Bridge in Bondsville is on River Street/State Street spanning the Swift
River between Belchertown and Palmer. It is a pin-connected Pratt Pony Truss bridge
with a span length of 77 ft. The bridge was built in 1880 by the Wrought Iron Bridge
Co. of Canton Ohio and has a number of unusual features, including patented ribbed
T lattice vertical members. The bridge has remained virtually intact since the time
of its original construction and therefore is historically accurate to the design of the
period (see Figure 11.8). A new bridge is currently being design by MHD for this
location. It is likely that dismantling will be performed during the summer of 2004.
We are first in line for ownership of this bridge from MHD for use on this Project.

BRIDGE NO. 8: WROUGHT IRON PRATT THROUGH TRUSS

The Call Road Bridge in Shattuckville, Ma. was constructed in 1895 by the Groton
Bridge & Manufacturing Company to span the North River. The bridge is a classic
example of one of the most common highway bridges of this era and is a 93 ft. span
six panel pin connected Pratt Through Truss bridge. (See Figure 11.9.)

FIGURE 11.8 Bridge No. 7 while still in service.

FIGURE 11.9 Bridge No. 8—Call Road Pratt through truss bridge.

Long Term Maintenance

Each bridge with be painted with an appropriate bridge paint as a part of the final reconstruction. Colors will be chosen that will enhance the location of the bridge. Bridge decking that requires minimal maintenance will be used. Each bridge will be inspected annually by a team of students from ASCE and the CEE Department. Any required maintenance will be performed by the CEE Department and ASCE students.

EXAMPLES OF STUDENT WORK TASKS

In addition to being involved in the various administrative and technical aspects of the work at various stages, students are also heavily in the fabrication of new components to replace components that are too heavily deteriorated to be used in the new construction. Figure 11.10 shows an example of a typical AutoCad drawing prepared by students showing the end connections and bearing plates for one of the bridges.

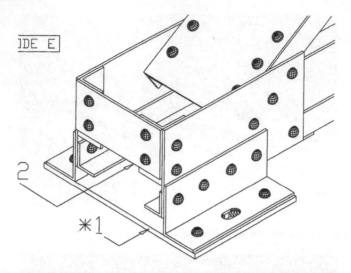

FIGURE 11.10 Typical AutoCad drawing of end connection and bearing plate.

FIGURE 11.11 Student fabricating a new truss plate in Engineering Machine Shop.

Figure 11.11 shows a student using a drill press in the Engineering Machine Shop to fabricate a new connection plate for a truss. Figure 11.12 shows students preparing two trusses for a trial assembly prior to moving the trusses and beams to the final location on campus. An initial assembly will be made of each bridge to ensure that all major components fit properly before they are transported to the site for final assembly.

Some tasks require the students to devise innovative solutions. For example, in order to remove some deteriorated components such as lower chord angles that had to be replaced, it was necessary to carefully remove existing rivets. In order to perform this work without damaging other components, students designed and built a portable hydraulic rivet press that could be used to press rivets out once the rivet head had bee ground flush with a small grinder. Figure 11.13 shows students using the press to remove rivets at a lower chord connection on Bridge No. 2.

FIGURE 11.12   Students performing trial assembly of trusses and beams.

FIGURE 11.13   Students using portable hydraulic rivet press to remove rivets.

SUMMARY: ANTICIPATED BENEFITS

It is envisioned that the Project will have a number of short-term and long-term benefits to the University, to outreach, and to the CEE undergraduate and graduate curricula. One of the goals of the Project will be to provide a unique opportunity to introduce concepts of Civil Engineering to the K-12 audience of students, with elementary and high school teachers encouraged to bring classes to the UMass campus to perform the walking tour. Figure 11.14 shows Bridge No. 1 in its final location spanning a small stream and serving as a pedestrian bridge on a sidewalk. The bridges will also be used in a number of CEE courses to illustrate fundamental concepts of mechanics and to discuss historical aspects of engineering design.

FIGURE 11.14   Bridge No. 1 at final location.

During the restoration and reconstruction, students will have the opportunity to participate in a number of typical construction activities that are important to all areas of CEE. The Project is being supported by the Vermont DOT, the MHD and is being coordinated with the UMass Office of Planning to develop locations to site the bridges. Most of the bridges will be reconfigured from their original design width to pedestrian width, while still preserving the original form and components to be historically accurate. Figure 11.15 shows the trial assembly of Bridge No. 2 which will be reconstruction at its final location in July of 2008. At the present, the Project is working with several groups to obtain three additional bridges. The project will save these bridges from destruction and will give them a new life for the next century.

FIGURE 11.15   Trial assembly of Bridge No. 2.

REFERENCES

Boothby, T. and Craig, R.J., 1997. Experimental Load Rating Study of a Historic Truss Bridge. Journal of Bridge Engineering, ASCE, Vol. 2, No. 1, pp. 18–26.

Green, P.S., Conner, R.J. and Higgins, C., 1999. Rehabilitation of a Nineteenth Century Cast and Wrought Iron Bridge. 1999 New Orleans Structures Conference, ASCE, pp. 259–262.

Schenk, T.S., Laman. J.A., and Boothby, T., 1999. Comparison of Experimental and Analytical Load-Rating Methodologies for a Pony-Truss Bridge. Transportation Research Record No. 1688, pp. 68–75.

Zuk, W. and McKeel, W.T., 1981. Adaptive Uses of Historic Metal Truss Bridges. Transportation Research Record No. 834, pp. 1–6.

12 Preservation of Stone Masonry Aqueducts on the Chesapeake and Ohio Canal

Denis J. McMullan and Douglas E. Bond

CONTENTS

HISTORY OF C&O CANAL AQUEDUCTS

The Chesapeake and Ohio Canal (C&O Canal), extending from Washington, DC to Cumberland, Maryland, is one of the most popular parks in the National Park System. Each year thousands of park visitors use the park's towpath to bike, hike, jog, and ride or otherwise use this park. But the use of this park is very dependent on towpath continuity. And towpath continuity is dependent on maintaining and rehabilitating the twelve aqueducts along the canal. Collapse of one or more of these historic aqueducts would severe the canal's towpath and would greatly limit the public's enjoyment of this park. This fact had led to intensive efforts to preserve and rehabilitate these 175-year old structures.

Construction of the 185 mile long Chesapeake and Ohio Canal began in 1828 in Georgetown, District of Columbia, and was intended to reach the Ohio River but was never completed beyond Cumberland, Maryland. The C&O Canal system included eleven stone aqueducts and one timber trough aqueduct, designed to carry the canal

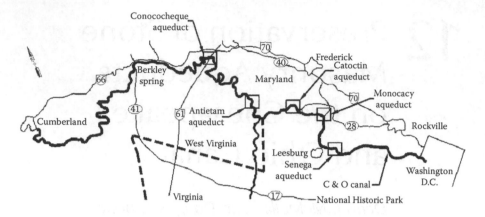

FIGURE 12.1   Map of the C&O Canal.

and boats across the major river tributaries that drain into the Potomac River along the canal's route.[2]

The C&O Canal depended on the Potomac River for its water supply which was both an advantage and a liability since the Potomac River is prone to severe flooding. The need to keep the level of the canal close to the level of the Potomac River and to keep the river tributaries navigable required careful attention to elevations and forced the designers to minimize the depth of the arch structures.

The Seneca Creek Aqueduct, designed by C&O Canal chief engineer Benjamin Wright, was the first aqueduct to be built on the canal. Construction commenced on November 27, 1828 and was completed in 1832. The aqueduct was a three-equal span segmented circular arch design. Each span was thirty three feet with a rise of seven feet eight inches. The west arch collapsed during a heavy flood in 1971 after which the National Park Service stabilized the structure by installing temporary steel beams across the missing span.[1]

The Monocacy Aqueduct was the second and largest of the eleven aqueducts erected along the canal. Also designed by Benjamin Wright it is often described by many historians as one of the finest canal features in the United States. This aqueduct is considered an icon of early American civil engineering. Its construction was begun in 1829 and was completed four years later in 1833. The aqueduct has six piers, two abutments, and seven, fifty-four foot arches, each with a rise of nine feet. The length of this aqueduct is 438 feet, and the total length of the structure including abutments is 516 feet.[8]

The Monocacy Aqueduct is sited at the mouth of the Monocacy River adjacent to the Potomac River. The aqueduct is frequently flooded, and is subjected to impact from debris that is washed against the structure on its upstream side. The National Park Service (NPS) had long been concerned about the structural stability of the aqueduct, and following the 1972 Hurricane Agnes flood, the Federal Highway Administration designed and installed internal grouted rods in the arch barrel and an external steel and wood banding system to temporarily stabilize the structure.

In June 1998, the National Trust for Historic Preservation identified the Monocacy Aqueduct as one of the eleven most endangered historic structures in the United

FIGURE 12.2   The Seneca Aqueduct.

FIGURE 12.3   The Monocacy Aqueduct with steel bracing and flood debris.

FIGURE 12.4 Completed stabilization of the Monocacy Aqueduct.

States. This led to a major construction effort in 2003/2004 which stabilized the aqueduct and enabled the obtrusive external steel banding to be removed.

Aqueduct number three is located at the Catoctin Creek and was constructed from 1832 to 1834. The stone masonry aqueduct was ninety two feet long between abutments and had three arches. The center arch was elliptical in form with a forty-foot span and ten-foot rise. Elliptical arches are rare among aqueducts. There are only four elliptical arches out of twenty two arches on the C&O Canal. They were most likely utilized to provide larger hydraulic opening but also possibly for aesthetic reasons. The two side arches were semicircular with a twenty-foot span and a ten-foot rise. The center elliptical arch had pronounced sag as early as the 1940's and probably earlier. The arch continued to sag until October 31, 1973 when it fell during a local flood and caused the consequent collapse of the west arch. The remaining east arch, wing walls, and east and west abutments remained standing but are vulnerable to further deterioration.[9]

Aqueduct number four is located at the mouth of the Antietam Creek. Built in 1834 it is 140 feet long and has three elliptical arches. The parapet walls were partially destroyed during the Civil War and then repaired in-kind.[5] The towpath parapet wall has deteriorated over the last several years with many displaced stones. Efforts are underway by the NPS to stabilize the structure.

The fifth aqueduct to be built by the C&O Canal Company was the three span Conococheague Creek Aqueduct. This aqueduct was also damaged during the Civil War with both Union and Confederate troops attempting to unsuccessfully destroy it. In the spring of 1865, the berm or upstream side of the aqueduct fell into the Conococheague Creek, briefly halting travel on the Canal. The cause of the collapse

FIGURE 12.5   The Catoctin Aqueduct.

FIGURE 12.6   The Antietam Aqueduct.

FIGURE 12.7   Wooden Wall Repairs at the Conococheague Aqueduct.

FIGURE 12.8   The Conococheague Aqueduct.

was believed to be the cumulative result of freezing and thawing coupled with the effect of damage during the Civil War. The wall was soon fixed, with a "wooden trunk", which was subsequently rebuilt with stone in 1870.

In 1920, this rebuilt stone parapet also collapsed and was replaced with a wood sheet wall supported on cantilevered timber beams set into concrete on the prism floor, which only lasted a few more years.

The arches of the remaining aqueducts, 6 through 11, are mostly intact although one is supported by steel bracing. The loss of the berm parapet and spandrel wall was a common failure for the C&O canal aqueducts. Of the 11 stone aqueducts, seven no longer have the berm parapet and upstream spandrel wall.

Disastrous floods and storms have been a part of the history of the C&O Canal since its very inception. During some storms, such as the giant flood of 1889, the Potomac River crested at 44 feet above the low-water mark, which would have over-topped all aqueducts in the area.

Damage from flooding in 1924 caused the abandonment of the canal which by then was owned by the Baltimore and Ohio (B&O) Railroad.

CONSTRUCTION/TECHNOLOGY

The designers of the C&O Canal aqueducts faced the challenges of building durable, watertight structures that would provide adequate clearances over the Potomac's major tributaries and yet maintain an elevation for the canal that could use gravity feed from the Potomac River. The structures would need to be robust enough to with-stand frequent flooding from the Potomac River together with often severe winters and the associated internal expansive forces from ice build up.

The foundations needed to withstand scouring forces from the river and be rigid enough to prevent settlement of the piers and abutments.

Soil borings have indicated that the piers and abutments were usually founded on relatively solid rock that was close to the surface. Underwater investigations have generally revealed little to moderate erosion of the rock at the interface with the foundation stones. This is supported by very few instances of significant settlement problems. The only known significant foundation problem occurred at the west pier of the Catoctin Aqueduct.

Stone for most of the aqueducts was obtained locally but in some instances stone was obtained a considerable distance from the aqueduct. For example, granitite for the Catoctin Aqueduct was transported by the B&O Railroad from Ellicott Mills Quarry near Baltimore. The Antietam Aqueduct is constructed of Tomstown Dolomite from a quarry three quarters of a mile to the east; the Conococheague Aqueduct uses limestone cut from a quarry three miles away.

The quality of local stone was often a matter of dispute. The initial construction of the piers for the Monocacy Aqueduct used stone from Nelson's Quarry located at nearby Sugar Loaf Mountain, four miles east of the aqueduct. However this stone turned out to be of such poor quality that the contractor was forced to dismantle the first three piers and rebuild them using a harder quartzite stone from Johnson's Quarry approximately halfway between the aqueduct and Nelson's quarry.[4]

The discovery and use of natural cement, also known as "hydraulic cement", that sets under water made the construction of shallow watertight arch structures feasible on the C&O Canal. Other earlier canals relied on a thick clay layer between the prism floor and the top of the arch barrel for waterproofing. This resulted in a greater height of spandrel wall between the top of the arch and the water table and short span heavy structures such as the five span aqueduct, over the River Inny (ca 1700's), on the Royal Canal Extension in Ireland.

FIGURE 12.9 McMullan & Associates engineer checking ice on arch soffit of the Monocacy Aqueduct.

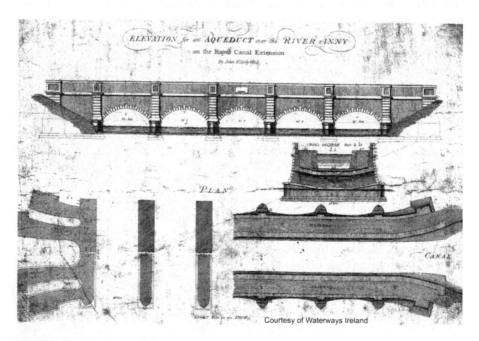

Courtesy of Waterways Ireland

FIGURE 12.10 Greater heights of spandrel walls at the Aqueduct over the River Inny, Ireland.

FIGURE 12.11   Interior mortared stone fill at the Conococheague Aqueduct.

Natural cement is made from naturally occurring limestone with appropriate argillaceous properties. It was therefore important for the early C&O engineers to find suitable limestone on the Potomac Valley. They conducted field testing of local limestone deposits to identify good candidates for the production of natural cements. Botelor's Mill, located immediately south of Shepherdstown, was the first natural cement mill built in the Potomac Valley. It provided natural cement to the Monocacy Aqueduct and numerous other structures along the C&O Canal. After completion of the canal this industry continued. Eleven cement mills were eventually constructed to produce natural cement. The Round Top Cement Mill west of Hancock, Maryland was one of the largest.[3]

Once the foundation stones had been laid, most likely inside timber cofferdams, the piers and abutments were brought up in rough cut stone faced with solid cut stone to the springline. Above the springline, the finish of the exterior stones was a higher quality. The solid cut stone in the piers stopped at the intersection of the extrados of the arches. The triangular volume between the adjacent arches was filled with mortared stone fill as can be seen in the exposed pier at the Conococheague Aqueduct.

The arch geometry was formed in wood planking on timber centering that was removed upon completion of the structure.

The semicircular arch that occurs in a few locations on the C&O Canal, with a rise to span ratio of 1:2 is the strongest shape of the arches used. However, this form results in short spans with numerous and expensive piers. The segmental circular arch was very commonly used on the C&O Canal with rise to span ratios varying from 1:4 to 1:6 for the Tonoloway Aqueduct and the Monocacy Aqueduct respectively. This shape provided a more efficient use of materials, longer spans, and sufficient hydraulic openings for high water conditions. In a few locations, namely at the Antietam and the Catoctin Aqueduct, elliptical arches were employed.

Aqueduct Arch Dimensions			
Aqueduct	Span	Rise	Rise to Span
Seneca	33	8	1 : 4.1
Monocacy	54	9	1 : 6
Catoctin (left/right)	20	10	1 : 2
Catoctin (center)	40	10	1 : 4
Antietam (left/right)	28	7	1 : 4
Antietam (center)	40	7	1 : 5.7
Conocheague	60	15	1 : 4
Licking creek	70	14	1 : 5
Tonoloway creek	80	20	1 : 4
Sideling hill	70	12	1 : 5.8
Fifteen mile	50	12	1 : 4.2
Town creek	60	15	1 : 4
Evitts creek	70	14	1 : 5

FIGURE 12.12   The aqueduct rise/span ratios.

A lot of attention was paid to the detailing of the ring stones (voussoirs) and the keystones. At the Catoctin Aqueduct the voussoirs have a margin around the four sides and a raised rock face finish. Voussoirs varied in height with the maximum at the springline and tapering to a minimum at the crown. This was applied even to the smaller circular arches on the Catoctin Aqueduct.

After the arch barrel was laid the spandrel walls were constructed on the voussoirs in a repetitive ashlar pattern. Spandrel stones were twelve inches to eighteen inches in depth with a regular pattern of header stones roughly four feet deep tying the spandrel stones to the stone fill.

The stone fill often referred to as "rubble fill" was actually carefully laid up large and small stones with mortared joints. After a section was laid for the day, hydraulic cement grout was poured into any small voids or holes left in the fill.

Once the stone fill and the spandrel walls had reached the height of the bottom of the prism, a decorative water table stone was set in the spandrel walls. The interior wall face stones of the towpath and berm parapets were started on the mortared stone fill and each of the four walls was carried up another six to seven feet to provide parapets that contained the waterway. The same mortared stone fill was used between the parapet walls. This was then covered by large twelve inch thick coping stones, usually six feet by three feet that cantilevered six to ten inches over the spandrel wall.

A decorative wrought iron railing was installed on the towpath parapet along the river side and a wooden mule guide rail installed on the canal side. Timber rub rails were also installed on the inside face of the towpath walls to protect the boats.

Although the builders made every effort to ensure a water tight structure, it was a very difficult task. The aqueduct prisms constantly leaked. On the C & O Canal several different methods of waterproofing were tried. At the Conococheague Aqueduct, the prism floor was overlaid with hard burnt brick laid on one edge on a bed of mortar one inch deep. Cement grout was then poured over the bricks to fill any gaps and provide an additional layer of protection.[6] On the Monocacy Aqueduct,

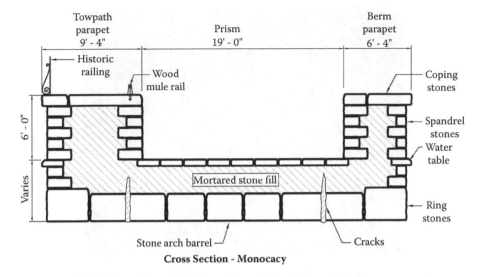

FIGURE 12.13 Cross section of the Monocacy Aqueduct.

the photographic and other historic documents strongly suggests that two inch wood planking was used. At the Catoctin Aqueduct, there is a two inch deep and one inch wide slot on the inside face of the parapet walls at the prism floor elevation suggesting that wood planking was also used here as a waterproof barrier.

After a collapse of the Catoctin Aqueduct berm parapet, and during the rebuilding effort, Chief Engineer Fisk in 1838 decided to use a new product "American Cement" patented by Thomas C. Coyle. Nine hundred and twenty four barrels of this cement were used in the reconstruction. This product contained resin and tar, and must have been applied hot as there were costs for the kettles noted.[10] Test pits in the prism revealed a layer of this 'resin cement' at the floor level.

In some locations, deteriorated stone masonry was replaced with Portland-Cement based concrete. At the Monocacy Aqueduct, a section of the berm parapet was rebuilt with concrete. Concrete was also used to fill voids and cracks. There was one crack in the arch barrel under the berm parapet at the Monocacy Aqueduct that appears to have been filled from above, probably by removing a section of the berm and pouring the concrete into the open crack. Concrete was also used to repair voids or deteriorated foundations as occurred at Pier #6 at the Monocacy Aqueduct.

To limit displacement of the coping stones, iron cramps were inserted into recesses in the surfaces of the coping stones to tie the stones together. At the Monocacy Aqueduct, in addition to the iron cramps, diamond shaped iron pins between the coping stones were used to limit differential lateral movements of the coping stones.

STRUCTURAL INVESTIGATIONS OF THE AQUEDUCTS

MECHANISMS OF DETERIORATION AND FAILURE

The strength of the masonry arches on the canal was dependent on several factors primarily geometry, stone and mortar quality and workmanship.

FIGURE 12.14   Iron cramps between coping stones at the Monocacy Aqueduct.

FIGURE 12.15   Hinge formations at the Catoctin Aqueduct.

The arch geometry and configuration can complicate other mechanisms due to the higher stresses caused by low rise to span ratios, thin arch barrels, and elliptical geometry. This was evidenced by the vertical sagging and the formation of hinges observed in the Catoctin Aqueduct prior to its collapse and at the center elliptical arch at the Antietam Aqueduct. Analysis of both of these elliptical arches indicates that with supports only at the springings, the line of thrust would be located outside the arch barrel stones causing the formation of hinges. However a stiff backing was provided by the mortared stone fill that is located behind the arch stones and this contributed to the resistance of horizontal arch thrusts and initially prevented the formation of hinges.

FIGURE 12.16   Underwater inspection of the remains of the west pier at the Catoctin Aqueduct.

Erosion of the rock upon which the pier foundations bear, created voids under the stones. The presence of these voids at the outside face of the foundation stones was observed during underwater investigations at the Monocacy and the Catoctin Aqueducts. The formation of voids in the foundation stones themselves as seen at the Monocacy Aqueduct in Pier No. 3 and the West Pier at the Catoctin Aqueduct may be linked to the erosion of the supporting rock.

There were many forces and conditions that acted upon the aqueducts that led to their deterioration and failures. The most basic of these is the lateral pressures on the parapets and spandrel walls from water in the prism and from saturation of the fill. The impact of canal boats on the parapets also imparted lateral forces on the parapets and supporting arch barrels. The berm parapet wall at the Conococheague Aqueduct collapsed when struck by a canal boat.[7] Lateral forces on the parapets would have been resisted by friction forces between the stones in the arch barrels and mortared stone fill. Since the arch barrel stones contained compression forces in the direction of span, there would have been considerable friction forces between the stones to resist lateral forces. However, as the constant water saturation and flow into the fill gradually caused the mortar to deteriorate, the head mortar joints between the arch barrel stones would open or the stones would crack. These are believed to have developed into the long longitudinal cracks that are common among the C&O Canal Aqueducts.

Periodic flooding of the streams and creeks caused saturation of the entire structure and allowed the intrusion of flood borne silt and debris into voids in the masonry. Several inches of silt were found behind the pier face stones at the West Abutment of the Monocacy Aqueduct, at the location of a large bulge.

FIGURE 12.17 Bulge at the Monocacy Aqueduct West Abutment.

Also, acidity of flood waters may have caused a gradual deterioration of the mortar. At the Monocacy Aqueduct, some of the mortared stone fill was found to have a dense sand material between the stones instead of a hard mortar. Testing of the sand determined that it contained elements of deteriorated hydraulic lime mortar.

The lack of water-tightness of the aqueduct prism itself led to other problems. Once the flow of water into and through the structure occurred, a mechanism for the loss of fines in the mortar and stone fill was established. Voids were created that weakened the mortared stone masonry fill and created a path for increased water flow. Evidence of the flow of water through the structures can be seen in historic photos of the Monocacy Aqueduct.

Freeze-thaw action of water in saturated masonry also contributed to the displacement of the stones and deterioration of the mortar between the stones. The C&O Canal Aqueducts exhibit much more movement than do other arch barrel canal structures that do not carry water, such as the aqueduct at the Schoharie Creek Aqueduct on the Erie Canal that used a wooden trunk to carry water and stone arches to support the towpath.

PHYSICAL INDICATIONS OF STRUCTURAL MOVEMENTS AND DETERIORATION

Prior to the design of rehabilitation measures, a detailed inspection and assessment was made of the aqueducts. There were many indications of deterioration and

FIGURE 12.18   Water flow through the Monocacy Aqueduct.

displacements noted. At the Monocacy and the Antietam Aqueducts, many of the coping stones that cap the parapets had tilted. In all of the aqueducts, several of the spandrel wall stones had slid on the top of the arch barrel ring stones, in some cases several inches.

There are many longitudinal cracks in the arch barrels and in some cases separation of the arch barrel stones by several inches. The cracks were a combination of a separation of the head joints between the arch barrel stones and cracks through the stones themselves. The cracks generally coincide with the width of the parapets and also are a greater magnitude under the berm parapets. There are also a few vertically displaced arch stones and missing arch barrel stones that may have previously dropped. In general the bed joints between the arch stones, those joints which transmit the principle arch forces in the direction of the span, were found in relatively good condition. In contrast, the head joints which are perpendicular to the bed joints were found to have missing mortar in many of the joints.

At the elliptical arches of the Catoctin and the Antietam Aqueducts, vertical deflection of the arch barrels is evident. This may have occurred immediately after removal of the centering when the arch thrust caused some lateral displacement until enough passive pressure was developed in the backing fill behind the arch. The deflection also may have been caused by a loss of backing material through erosion and the creation of voids so that the horizontal thrust was no longer resisted by the backing but only by the masonry foundation at the springline. Analysis indicated

FIGURE 12.19   Longitudinal cracks in one of the Monocacy Aqueduct's arch barrels.

that this shift or lowering of the resistance point causes tension in the arch that would allow hinges to form and lead to the deflection of the arch.

Voids in the mortared stone fill were found at the base of test pits in the prism of the Monocacy Aqueduct, under the coping stones, and above the cracks in the arch barrels. Water testing of the voids under the coping stones indicated that the cracks and voids in the fill extended from the top of the parapets through the arch barrel. The voids were probably created by the loss of interior fill material through the large cracks in the arch barrel.

The iron railings also contributed to deterioration of the structure. The iron railing was set in holes drilled into the coping stones and about six inches from the edge of the stone, lining up with the face of the spandrel walls. The posts were placed every eight inches on center and were set in lead. Due to a combination of rust and water expansion in the hole, the coping stones often cracked along this line and the edges of the coping stones were often lost. The canal company made repairs by setting the railings back on metal straps wrapped over the parapets.

During flood events, the wrought iron railings on the towpath often caught debris which led to the coping stones being pulled off the parapet into the river below.

ANALYSIS OF EXISTING CONDITIONS

A load rating of the arches for the Monocacy and the Antietam Aqueducts was performed to determine if the existing arch barrels could safely support vehicular maintenance traffic. The method used was derived from a procedure widely used in the United Kingdom (UK) for evaluating existing arch bridges in various states of disrepair. This method, initially developed by the Military Engineering Experimental Establishment (MEXE), provides for applying condition factors to the strengths of the stone and mortar used in the analysis.

To evaluate the arches, a finite element model was developed to determine the forces in the arch barrel on a per unit length basis. Loads from the fill, the parapet, and vehicles were distributed along the length of the arch and the width of the arch barrel. Stone strengths were derived from compression testing and mortar strengths assumed from published values of hydraulic cement mortars. The UK method provides for determination of the masonry strength as a function of the stone and mortar strength and sizes. Other existing conditions which affected the analysis were taken into account including the longitudinal cracks which separate the arch barrel and missing arch barrel stones. At the Monocacy Aqueduct, the arch barrels were determined to marginally support an H-15 vehicular loading, as defined by the American Association of State Highway and Transportation Officials, without strengthening. Due to the missing stones at the Antietam Aqueduct, the analysis indicated that the arch barrels need repair in order to safely support an H-15 loading.

At the Catoctin and the Antietam Aqueducts, there is an eccentricity on the pier foundations due to the unbalanced vertical and horizontal components of the smaller circular arch on one side and the larger elliptical arch on the other. The magnitude of the loads along the width of the piers varies from a maximum at the face of the spandrel walls to a minimum at the center of the prism due to the parapets which contribute more to the load than the arch and fill self weight. The weight of the water in the prism would have somewhat offset this variation in load in the pier along its length. At the Catoctin Aqueduct, a finite element analysis of the stone arches indicated that the unbalanced horizontal components produced a resultant 76 kip vertical load located outside the middle third of the base of the pier foundation due to the height of the pier. A resultant force that acts outside the middle third of a foundation creates net uplift on one edge of the foundation. Assuming that the pier stone masonry cannot resist tension, this load results in a maximum bearing stress at the base of the pier of nearly 45 ksf, a relatively high value. By reconstructing the arch barrel with a material that does not rely on arch action for stability and has a tension resisting capability, such as reinforced concrete, this bearing pressure would be reduced.

STABILIZATION AND RESTORATION TECHNIQUES

STABILIZATION OF THE MONOCACY AQUEDUCT

After the flood of 1972, concern about the longitudinal cracks in the arch barrels of the Monocacy Aqueduct led the Federal Highways Administration in conjunction with the National Park Service to implement emergency measures to prevent a

FIGURE 12.20   Steel banding installed in the 1970's at the Monocacy Aqueduct.

collapse. Grouted reinforcing bars were installed into the arch ring stones on each side and extended across sixty percent% of the width of the arch barrel. The bars were drilled at a slight angle to the horizontal.

In addition, the entire aqueduct was banded with exterior steel rods and wood blocking to prevent further lateral movements of the parapets and spandrel walls. In the 1990's, a study by McMullan and Associates determined that the wood blocking had deteriorated and the rods were no longer tensioned such that the banding was ineffective. The lack of tension in the rods was also later confirmed by an independent study by the American Society of Civil Engineers (ASCE). The banding detracted from the historical appearance of the structure and was an obstacle to visitors. As a result, alternatives for removing the banding and using other means of stabilizing the structure were developed and a stabilization design by McMullan and Associates was completed in 2004.

In order to rely on the grouted reinforcing bars as an effective part of the stabilization work, it was necessary to check the condition of the reinforcing bars installed as part of the FHWA effort. This was accomplished by measuring the corrosion rate of the bars. The end of several bars in the core holes were exposed and bars electrically connected to a copper-copper sulfate portable half-cell that measured the potential in volts of the reinforcing bars. The location of the most negative potential on the bars was selected for the corrosion rate measurement. Of the twenty two bars tested, twenty one were found to be in the passive range with no corrosion, and one was found to be in the low corrosion range. A visual survey of the arches found no rust stains on the arch soffits. In addition, grout covering the bars was observed in some of the open cracks. This confirmed that these bars were still in good condition and could be used as lower tension ties.

An upper level tension tie for the parapets was accomplished by constructing a reinforced concrete slab with reinforcing bars drilled and grouted into the spandrel wall face stones.

FIGURE 12.21   Drilling hole for reinforcing bars into the parapet at the Monocacy Aqueduct.

The surface of the slab was sloped to promote water run-off and a topping slab introduced on the reinforced slab that had a scored finish to simulate a wood plank bottom.

The aqueduct arch barrels were repaired by a combination of pointing and grouting. After cleaning the joints with pressurized water, the joints were pointed using a cement lime mortar similar to a Type N mortar. Different crack repair methods were used for different ranges of crack widths with the widest crack being eight inches. At the location of large cracks or cavities, tubes were inserted into the cracks and cavities and sealed with mortar. Then a cementitious grout was pumped into the cavities.

The coping stones on the top of the parapets were removed and large pockets of silt and voids in the mortared stone fill were exposed. The silt was removed by hand excavation and the use of vacuums. In some cases water flushing was used to remove the silt through drill holes or cracks. The surfaces of the stones were pointed using mortar. A cementitious grout was placed into the voids and cracks and allowed

FIGURE 12.22   Flow grouting the prism floor at the Monocacy Aqueduct.

to seep into the fill. Even with the pointing, grout leakage did occur as expected. By placing barges with scaffolding under the arches during grouting, the leakages were collected and addressed by driving in wedges at the location of leaks. Admixtures were used to promote the flow of the grout, but also caused the grout to become viscous once the initial velocity of the grout flow slowed, this significantly helped to limit leakage from the aqueduct.

In addition to the work on the parapets, the earth fill on the prism floor was excavated to expose the mortared stone fill and voids and cracks between the stones. A cementitious flowable fill was placed on the exposed mortared stone fill to flow into the cracks and voids and provided a base for the reinforced concrete slab.

The areas of spandrel and pier face stone bulging were stabilized by pinning the stones back using drilled and grouted reinforcing bars. First, cores were taken at the drill hole locations and the core plugs removed and saved. The holes were then drilled and reinforcing bars inserted and grouted. The cores were then used to cover the drill holes. Stone shavings from the coring were added to the mortar mix used to fasten the core plugs. This resulted in minimal visual impact to the face stones.

To prevent flood debris causing damage to the railing and coping stones, a removable railing was introduced, similar in appearance to the historical railing. The posts for the railing were inserted into stainless steel sleeves that were grouted into the coping stones. The railing sections overlap so that only one section needs to be unbolted to remove the whole length of railing. The single bolt is secured with a padlock. The NPS monitors flood activity and has made the removal of the railings

FIGURE 12.23   New and existing railing at the Monocacy Aqueduct.

part of their standard procedures for flood preparations. In a mock drill the Park were able to remove the entire railing in under one hour.

Floor drains were introduced along the five hundred foot plus length of the Monocacy Aqueduct, at the center of each arch barrel to ensure rapid removal of water from the aqueduct prism.

RECONSTRUCTION OF THE CATOCTIN AQUEDUCT

The reconstruction of the Catoctin Aqueduct (planned for 2008/9) will use a mix of new and original aqueduct stones that had been salvaged and buried by the NPS after the collapse of the arches. Except for a few historical photographs there is very little documentation of the original arch geometry or elevations available.

Underwater investigation revealed voids at the interface between bedrock and some of the foundation stones. Forms will be set along the foundation and concrete pumped into the voids. At the West Pier, there is a hole in one side of the pier eight feet wide and almost five feet deep, filled with creek sediment and debris. Formwork will be placed around the entire pier, sealed, and concrete placed into the hole. In addition, reinforcing bars will be drilled and grouted through the perimeter stones of the remaining pier and grouted into the bedrock. The bars will be developed for compression and tension loading to account for eccentricities on the pier. The grout will be pumped with low pressure with the intent of filling voids between the foundation stones and encasing the reinforcing bars.

The salvaged aqueduct stones excavated by the NPS proved to be a mix of different types of stones including copings, parapet wall, spandrel wall, water table, arch barrel, and arch ring stones. Many of these stones have small but significant differences between them and identification of the stones proved difficult. The NPS sorted the stones into piles on wood blocking based on their type. They also numbered and measured the dimensions of the coping stones. Using AutoCAD, these stones were put together as a puzzle so that the entire towpath surface could be reconstructed.

FIGURE 12.24   The Catoctin Aqueduct arch stone dimensions.

FIGURE 12.25   The device to measure arch stones at the Catoctin Aqueduct.

Matching up the recesses cut into the coping stones for placement of iron cramps and broken edges of stones was key to solving the puzzle.

Identifying the exact location of the arch ring stones proved to be more challenging. The NPS recovered only enough ring stones for part of the downstream side of the aqueduct. The ring stone dimensions vary in depth, width and height.

The arch had to be re-created in order for the proper placement of stones to result in a smooth curve. Historically, the layout of elliptical arch stones has been done using graphical methods that proved impossible to duplicate from the historical photographs. Instead, The NPS engineered a simple device for measuring the stones which have tapered sides and a curved top and bottom.

These measurements were used to create an AutoCAD block of each stone. In addition, numbered foam templates of the stones were created which the NPS used

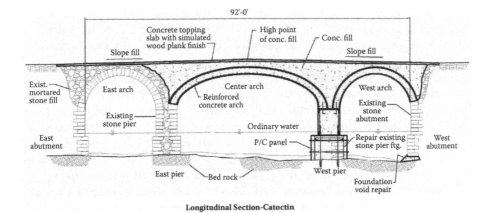

FIGURE 12.26   Longitudinal section of the Catoctin Aqueduct.

to lay out the arch and determine the placement of the stones. Through these methods, an AutoCAD drawing of the arch was created and the location and dimensions of several missing voussoirs have been identified.

Rather than attempt to place each spandrel stone in its original position, the NPS decided that an acceptable alternative was to replicate the stone pattern of the original spandrel wall. The original stones recovered by the NPS will be used to the extent possible and gaps filled in with new stone similar in color and finish to the original. The stones from the inside faces of the parapets were differentiated from those on the outside spandrel walls by rub marks from canal boats or bolt holes. A New England granite has been identified as a good match to the original stone.

Reinforced cast in place concrete arch barrels will be used instead of reconstructing the original arch barrels. The arch ring stones will be cut to match the width of the spandrel wall stones and anchored to the face of the concrete arch barrel.

Above the arch barrel, a concrete fill will be used instead of a mortared stone fill. This will allow the introduction of transverse reinforcement within the fill. In addition, instead of relying on original tie stones that extended into the rubble stone walls to anchor the face stones of the spandrel wall, stainless steel bars will be drilled and epoxied to the back of the stones and extend into the concrete fill.

The top of the concrete fill will be sloped from the center of the aqueduct to each end of the aqueduct at a two percent slope where site drains will carry water outside the canal prism. A concrete topping slab will be placed on the concrete fill with a scored surface to simulate a wood planked surface.

PRESERVATION FUNDING

The success of the Monocacy Aqueduct project was due to the efforts of many parties including the C&O Canal Association that raised over $200,000 to contribute to the project. There was strong local political and community support from State Senators and Congressional Representatives. It also sparked new public awareness of these important historic structures on the C&O Canal and the urgent need for their preservation.

At the Catoctin Aqueducts, the National Park Service is partnering with the Catoctin Aqueduct Restoration Fund, Inc. (CAR Fund) a 501 (c) (3) public charity formed in 2005 for the single purpose of raising funds for the restoration of the aqueduct.

The strong local fund raising effort has provided numerous educational opportunities that have further raised the awareness of the C&O Canal and its aqueducts to a new level, such that the Governor of Maryland has endorsed providing matching Transport Enhancement Program (TEP) funding for the project. Numerous local businesses and individuals have also contributed significantly to the fund raising.

With limited Federal Funds available for historic preservation, the future success of the C&O Canal Aqueduct preservation and restorations will be strongly linked to the success of private/public cooperation such as the NPS/CAR partnership.

Through a private-public partnership, the NPS is rehabilitating the historic aqueducts along the C&O Canal thus ensuring towpath continuity and continuing high usage of this very public park. McMullan & Associates, Inc. has been pleased to lend its structural engineering understanding and experience to this important undertaking.

REFERENCES

1. Fields, Thomas E. *Historic Structures Report Seneca Creek Aqueduct and Lock 24 (Riley's Lock)*, Chesapeake and Ohio Canal National Historical Park, 1979.
2. Hahn, Thomas F. *Tow Path Guide to the C&O Canal*, published by the American Canal and Transportation Center, 1988.
3. Hahn, Thomas, F.; Kemp, Emory L. *Cement Mill's Along the Potomac River* Institute for the History of Technology & Industrial Archeology Monograph Series Vol. 2 Number 1 – 1994.
4. Kapsch, Robert J. PhD; Kapsch, Elizabeth Perry. *Monocacy Aqueduct on the Chesapeake & Ohio Canal*, Medley Press, 2005.
5. Luzader, John F. *Historic Structures Report, The Antietam Aqueduct, Chesapeake and Ohio Canal National Historical Park*, 1964.
6. Luzader, John F. *Historic Structures Report Conococheague Aqueduct Chesapeake & Ohio Canal*, 1963.
7. Luzader, John F. *Historic Structures Report, The Conococheague Aqueduct*, Chesapeake and Ohio Canal, National Historical Park 1961.
8. Official National Park Handbook, *Chesapeake and Ohio Canal* by National Park Service.
9. Unrau, Harlan D. *Historic Structures Report, The Catoctin Aqueduct, Chesapeake and Ohio Canal National Historical Park*, 1976.
10. Unrau, Harlan D. *Historic Structure Report the Catoctin Aqueduct* Chesapeake and Ohio Canal National Historical Park, 1976.

13 Rehabilitation of Two Historic Timber Covered Bridges in Massachusetts

S. D. Daniel Lee and Brian Brenner

CONTENTS

INTRODUCTION

Timber covered bridges are difficult structures to rehabilitate, especially those that are registered in the National Register of Historic Places. Historical Commission members, at the state or local level, have a strong interest in preserving as much of the original fabric of the structure as possible. This goal can conflict with requirements to upgrade the structures to meet modern design criteria. Occasionally, compromises to the structural components of the bridge are made to increase the load carrying capacity of the structure to satisfy present day traffic loading. These bridges were originally designed to carry farm animals, horse buggies, and wagons and not for the truck loading design criteria currently specified by AASHTO. The original loads are much smaller than the much heavier loads from modern vehicles.

In our discussion, we will distinguish between "restoration" and "rehabilitation". Restoration is a process where a historic structure is repaired without necessarily increasing structural capacity from its original condition. Rehabilitation, however, involved not only repair, but a general upgrading of the structure to meet more current design criteria and loading requirements.

In this paper, we discuss rehabilitation of two timber covered bridges in Massachusetts. Both are well over a hundred years old and are listed in the National Register of Historic Places. The first bridge, the Burkeville Covered Bridge is in the Town of Conway, MA. It carries Main Poland Road over the South River. The second bridge, the Arthur A. Smith Covered Bridge is in the Town of Colrain, MA and it carries Lyonsville Road over the East Branch of the North River.

The owner of these two historic timber bridges is the Massachusetts Highway Department, MassHighway. MassHighway required both bridges to be rehabilitated to carry pedestrian loading, and also be capable of carrying AASHTO H 15-44 truck loading. During normal operation, the rehabilitated bridges were to be closed to vehicular traffic with removable bollards and chains. When needed for emergency use, authorities could remove the pad locked bollards and chains at the portals of the Arthur A. Smith Covered Bridge and removable chains at the Burkeville Covered Bridge.

The decision to primarily allow pedestrian use lies in the configuration of the existing railing and the width of these two structures. The Burkeville Covered Bridge has a single rectangular timber rail that is attached directly to the diagonals of each of the pair of Howe trusses. The Arthur A. Smith Covered Bridge has a more robust new timber railing system in comparison to the Burkeville Covered Bridge. However, neither bridge has a railing system that is strong enough to withstand present day design loading requirements from vehicular impacts.

Most significantly, the interior width of each of the two bridges is too narrow to accommodate an upgrade to a crash-tested timber bridge railing. In particular, the citizens and officials in the Town of Conway argued strenuously to open the rehabilitated Burkeville Covered Bridge to regular vehicular traffic, with a posting for 15-ton vehicles. However, based on concerns on the lack of crash-tested railings, which would protect both the historic structure and the motoring public, the decision to permit only regular pedestrian traffic remained in effect.

APPROACH

The structural integrity of the two bridge structures had to be carefully evaluated. MassHighway procured the services of a design team consisting of Fay, Spofford & Thorndike, LLC (FST), a design consulting firm, and Wood Advisory Services, Inc. (WAS), a timber specialty consulting firm. The team had to develop a strategy to evaluate the load carrying capacity of each member of the timber structure in order to determine what modifications would be needed to provide for the upgrade in load carrying capacity.

Since very little usable information was available for the two bridge structures, specific measurements, species of the timber components, and overall details and physical conditions of the member components had to be gathered from each of the two bridge sites.

A number of field inspection and data gathering trips were made by FST to acquire field measurements of the timber structures and theirs component parts for computer model analyses and evaluation.

Using basic wood science technology and laboratory analyses, along with simple common tools, such as a hammer and a pick, WAS performed assessments on the condition of the existing timber members of each bridge structure. A small diameter resistance drill with plotting capability was used to sample both exposed surfaces, as well as hidden areas of the timber members, with a goal of evaluating internal defects in the timber members, such as dry rots and insect infestation.

WAS also performed visual and laboratory identification of the species of the timber members and visual classification of the grading for the main load carrying members of the structural members and components of the structural system. These determinations and species identifications were used by FST to calculate member properties for computer model analyses and evaluation.

DESIGN METHODOLOGY

Old timber structures have varying amounts of construction tolerance for connections and member lengths. To further complicate the design analysis, member properties are not uniform throughout the length of the members or among members. Lengths of symmetrical members can vary because the wood was handcrafted. Seasonal variations in moisture content in each member components also affect the stress distribution in the structural system. Overall, the structural system of the two bridges does not behave elastically from the standpoint of a precise computer model analysis.

For these reasons, a three-dimensional computer analysis model could not precisely predict the member forces in the timber structure for any given loading combination, due to the imprecise nature of the structural system. This is true, of course, for any structural model, but the imprecision introduced by the old timber members led to additional concerns as compared to a model for a new bridge made out of steel. However, the resultant design forces and stress levels in the member components of the system, according to the model analyses, were evaluated and member components modified for design with sufficient factors of safety. Pin connected timber structural systems are resilient and have the ability to redistribute member forces. The ability for localized deformation of wood cells contributes to the characteristic feature for timber systems to redistribute large localized load into other members of the structural system, as long as there is sufficient redundancy. This capacity for load redistribution helped to enable analyses and designs leading to a satisfactory rehabilitation of the two subject covered bridges. This systematic resiliency is not easily modeled, and it must be evaluated carefully based on experience and taking into consideration the particular features of the structural systems being evaluated.

THE BURKEVILLE COVERED BRIDGE

DESCRIPTIONS OF STRUCTURE

The Burkeville Covered Bridge was built in 1871 (see Figure 13.1). The superstructure is approximately 106 feet long by 17 feet wide with a 12-foot 10-inch wide lane. It is supported on a pair of stone masonry abutments. Past repairs had covered both abutments with a solid smooth concrete facing.

The structural system is comprised of a pair of Howe trusses (with 8 panels each), 9 floor beams, longitudinal stringers, and transverse deck planks.

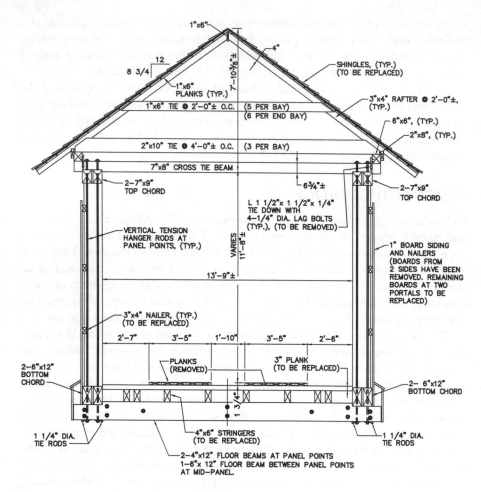

TYPICAL SECTION − EXISTING

FIGURE 13.1 Cross section of existing bridge structure.

Each Howe truss has 5 pairs of opposing timber diagonals arranged symmetri-
cally about the centerline of the span. Cross section of the diagonals ranges from
10×14, the largest, at the end and reduces in size toward mid-span from 8×12,
7×12, 7×12, and 6×12, the smallest.

Vertical members of the truss are made up of pairs of steel hanger rods at each
panel point of the truss. There are 9 pairs of hanger rods per truss. The 2 pairs of
hanger rods at and from the ends of the truss are 1-½ inches in diameter. The remaining 5 pairs of interior panel point hanger rods are 1-1/4 inches in diameter.

The top chord is made up of a pair of 7×9 timbers, stitched together with steel
bolts and intermittently spaced timber block shear keys. A space is created between
the pair of 7×9's with the intermittent shear keys. This space served to form the width
of the two-component top chord to receive the 12-inch wide timber diagonal members. Compression splices for the 7×9's are staggered between the pair of 7×9's.

Similar to the top chord, the bottom chord is made up of a pair of 6×12 timbers, stitched together with steel bolts and intermittently spaced timber block shear keys. However, each staggered tension splice is made with a pair of steel plates with multiple transverse square steel shear keys that are dovetailed precisely with the timber members to form a very tight splice. Many of these tension splices were meticulously formed between the steel shear keys and the dovetails in the timber.

A pair of longitudinal 1 ¼-inch diameter steel rods is located at the bottom of each bottom chord, mounted to the bottom of the floor beams. The two ends of these steel rods are bent upward and mounted with a steel plate that bears against the ends of the bottom chord. It is interesting to note that the bends in the rods are not restrained. Therefore, as the 2×12's of the bottom chord elongate from tension loading, the pair of steel rods tend to straighten out a bit at the bend points, thus providing very little tension capacity to assist the bottom chord.

There are 9 original floor beams at each of the panel points, supported by the pairs of vertical steel hangers. These floor beams are positioned underneath the bottom chords. Each floor beam consists of two 4×12's, with the steel hanger rods sandwiched in between. The pair of 4×12's is stitched together with 12 steel bolts.

In addition to the floor beams at the panel points, there are 10 mid-panel floor beams that are positioned under the bottom chords, midway between panel points. These mid-panel floor beams are single 6×12 timbers and they are attached to the bottom of the bottom chords with a pair of steel bolts at each connection point. It is not clear that these mid-panel floor beams were part of the original design. The members may have been added to improve load carrying capacity, when the timber decking was replaced. (As part of the current rehabilitation, these mid-panel floor beams have been made inactive, but preserved as part of the historic fabric of the timber structure.)

The stringers are 4×6 timbers, supported by panel point floor beams and mid-panel floor beams. There are six stringers spread across the top of the floor beams supporting transversely placed 3-inch thick timber planks.

For lateral stability of the covered bridge, timber cross braces are provided between the pair of bottom chords and cross tie beams that are on top of the top chords at each panel point. These 7×8 top chord cross tie beams also support the pair of steel hanger rods at each panel point. There is no additional lateral bracing at the bottom chord level other than diaphragm action provided by the 3-inch thick timber decking.

The gabled roof is framed by 3×4 timber rafters, spaced at approximately 2-foot on center. 1×6 timber planks provide support for the cedar shingle roofing. Two tiers of roof ties provide rigidity to the roof framing system. The upper tier of roof ties is made of 1×6's at approximately 2-foot on center. The lower tier of roof ties is made of 2×10's at approximately 4-foot on center. The entire roof framing is supported by the 2×10's, seated on top of a pair of 6×6 eave beams that are mounted on the ends of the cross tie beams. This fairly rigid roof framing provides additional lateral stability to the upper portion of the structural system.

DESIGN ISSUES AND MODIFICATIONS TO UPGRADE LOAD CARRYING CAPACITY

Capacity for two design loading conditions needed to be provided for in order to satisfy MassHighway's design requirements. The first condition was a design live

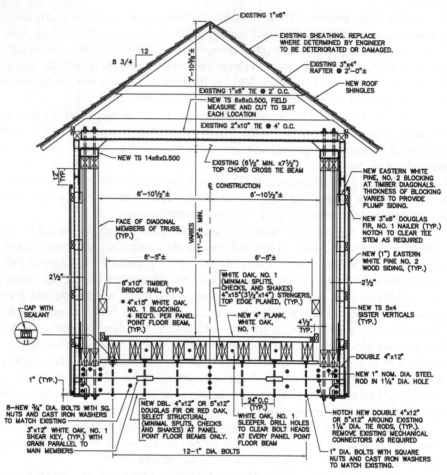

EXISTING 1"x6"

EXISTING SHEATHING. REPLACE
WHERE DETERMINED BY ENGINEER
TO BE DETERIORATED OR DAMAGED.

12
8 3/4

EXISTING 3"x4"
RAFTER @ 2'-0"±

NEW ROOF
SHINGLES

EXISTING 1"x6" TIE @ 2' O.C.

NEW TS 6x6x0.500, FIELD
MEASURE AND CUT TO SUIT
EACH LOCATION

EXISTING 2"x10" TIE @ 4' O.C.

NEW TS 14x6x0.500

EXISTING (8½" MIN. x7½")
TOP CHORD CROSS TIE BEAM

NEW EASTERN WHITE
PINE, NO. 2 BLOCKING
AT TIMBER DIAGONALS.
THICKNESS OF BLOCKING
VARIES TO PROVIDE
PLUMP SIDING.

12"
TYP.

€ CONSTRUCTION

6'-10½"±   6'-10½"±

FACE OF DIAGONAL
MEMBERS OF TRUSS,
(TYP.)

VARIES
11'-5"± MIN.

NEW 3"x6" DOUGLAS
FIR, NO. 1 NAILER (TYP.)
NOTCH TO CLEAR TEE
STEM AS REQUIRED

6'-5"±   6'-5"±

NEW (1") EASTERN
WHITE PINE NO. 2
WOOD SIDING, (TYP.)

2½"

6"x10" TIMBER
BRIDGE RAIL, (TYP.)

WHITE OAK, NO. 1
(MINIMAL SPLITS,
CHECKS, AND SHAKES)
4"x15"(3½"x14") STRINGERS,
TOP EDGE PLANED, (TYP.)

2½"

CAP WITH
SEALANT

* 4"x15" WHITE OAK,
NO. 1 BLOCKING.
4 REQ'D. PER PANEL
POINT FLOOR BEAM,
(TYP.)

NEW 4" PLANK,
WHITE OAK,
NO. 1

4½"
TYP.

NEW TS 5x4
SISTER VERTICALS
(TYP.)

DOUBLE 4"x12"

1" (TYP.)

NEW 1" NOM. DIA. STEEL
ROD IN 1¼" DIA. HOLE

8—NEW ¾" DIA. BOLTS WITH SQ.
NUTS AND CAST IRON WASHERS
TO MATCH EXISTING

NEW DBL. 4"x12" OR 5"x12"
DOUGLAS FIR OR RED OAK,
SELECT STRUCTURAL,
(MINIMAL SPLITS, CHECKS
AND SHAKES) AT PANEL
POINT FLOOR BEAMS ONLY.

24" O.C.
(TYP.)

WHITE OAK, NO. 1
SLEEPER. DRILL HOLES
TO CLEAR BOLT HEADS
AT EVERY PANEL POINT
FLOOR BEAM

NOTCH NEW DOUBLE 4"x12"
OR 5"x12" AROUND EXISTING
1¼" DIA. TIE RODS, (TYP.).
REMOVE EXISTING MECHANICAL
CONNECTORS AS REQUIRED

3"x12" WHITE OAK, NO. 1
SHEAR KEY, (TYP.) WITH
GRAIN PARALLEL TO
MAIN MEMBERS

12—1" DIA. BOLTS

1" DIA. BOLTS WITH SQUARE
NUTS AND CAST IRON WASHERS
TO MATCH EXISTING.

TYPICAL SECTION – PROPOSED

FIGURE 13.2 Cross section of bridge structure with proposed modifications.

load of an AASHTO H 15-44 truck (without impact) or AASHTO pedestrian loading of 85 psf, with pedestrian loading as the predominant design loading condition. The second condition was a design wind load consistent with general practice. (See Figure 13.2.)

It was determined early in the design process that the AASHTO wind load would not be appropriate, because the covered bridge had an elevation profile more like a building. AASHTO wind loading criteria provides for a uniform pressure on exposed girder surfaces that conservatively does not take into account variation and reduction in wind pressure closer to the ground. However, wind pressures from a building code do specify variation in pressure with elevation. Therefore, it was decided that an appropriate, location-specific wind load, according to the state building code, would be appropriate. The covered bridge is not very different from a barn with opened doors at front and back. The design wind pressure used for design was 12 psf for the given locale.

Since the covered bridge has a gable roof, an appropriate location-specific snow loading was also added to the loading combinations for determination of maximum design forces in the structural support system.

From the design perspective, it was quite challenging to satisfy either one of these two specified design conditions for rehabilitation. Unlike a restoration project, to upgrade a timber structure for rehabilitation is much more difficult to execute. Given the disparity between the original design loading conditions and the current design requirements, most, if not all, of the existing structural timber members or components would have to be sistered in order to increase the load carrying capacity, or replaced with larger and stronger species of timber to meet the greater load demands. This approach may be successful in achieving the desired structural capacities, but the results may not be satisfactory from a historic preservation perspective.

During the rehabilitation design stage of this bridge, our first design attempt considered sistering most of the diagonals and the top chords. All of these are compression members, where sistering with additional timber elements can be achieved without too much difficulty.

The bottom chords were much more difficult to modify. The pair of tension bottom chords are detailed such that adding timber sister elements are not possible. To overcome the significant increase in tension stresses in the pair of 2 × 12's of each bottom chord, a post-tensioned high strength steel rod is added to relieve and minimize the tension stress in the 2 × 12's. The post-tensioning load is designed to keep the maximum tension stress in the pair of 2 × 12's within their allowable design value. This post-tensioned high strength steel rod is the secondary element selected to reinforce the critical tension bottom chord against accidental overloading with higher live loads than intended by design.

The existing timber framing had not been designed to resist much wind load. As can be seen from Figure 13.1, there are no knee braces in the cross section of the structure. One could argue that the pairs of vertical hanger rods can provide a small amount of resistance against lateral racking due to transverse wind loading. However, it could not be determined that they have sufficient capacity to resist the current design wind pressure of 12 psf.

An attempt was made to develop a triangular steel plate connection between the 7 × 8 cross tie beams and the 7 × 9 top chord members to provide lateral resistance. Each of these steel connection plates is to be covered with a timber facing. However, the connection was so unwieldy, that MassHighway took exception to the detailing. An alternative solution had to be found to satisfy the lateral wind load case.

After many discussions, a second method was developed to resolve the wind induced design forces. This second approach developed from discussions on resolving an unrelated detailing issue. The Historical Commission commented critically on the sistering elements for the top chords and diagonals. They wanted to see if there is another way to increase the load carrying capacity without increasing the size of the members. They reasoned that perhaps in the future, a new technology or method might be available to make the historic timber members stronger, restoring the members to their original size and configuration. If the members were sistered, there would be numerous disfiguring bolt holes remaining in the historic timber members

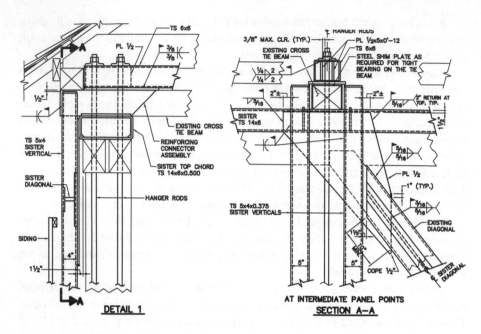

FIGURE 13.3   Details of steel supplemental frame at top chord panel point.

after removal of the sistering elements. Based on this consideration, we needed to develop a different method for strengthening.

One current philosophy in rehabilitation of historic structures is to add components that are distinctive, instead of trying to hide or mask the new structural elements. The approach then would be to create strengthening elements, components or members that are distinctive, yet harmonious, so that it is clear what their intended function is. One can easily tell them apart from the original components or fabrics of the historic structure.

Using this approach, instead of strengthening by using sistering elements, a second method of strengthening was developed. This time, a steel supplemental framing system was developed to compliment the top chords, cross tie beams, verticals, and diagonals. A strong moment connection is developed between the steel cross tie beams and the steel verticals to resist the transverse wind load.

To maximize load carrying capacity of the bottom chords, thrust restraint devices were added to engage the pair of existing 1 ¼-inch steel rods for their tension capacity.

Structural Analysis

To effectively analyze and evaluate the capacity of each structural member for the new design loading, a three-dimensional GT STRUDL computer model was created. Loading combinations with dead load; live load (without impact for timber structures)—H 15 truck or 85 psf pedestrian; wind load; and snow load were evaluated to determine maximum design forces in each component of the structural system.

(a)

FIGURE 13.4   Burkeville covered bridge before rehabilitation (a) and after rehabilitation (b and c).

The 3-D model comprised of the pair of Howe trusses, floor beams, new bottom lateral bracing at bottom chord level, cross tie beams, and top lateral bracing at top chord level.

As anticipated, design member forces were too high for the existing timber components of the trusses. The STRUDL model was modified to include steel components for the supplemental framing system. Ultimately, the existing timber framing and the new steel supplemental framing were designed to share the upgraded design loads according to each component's relative rigidity within the combined structural system of timber and steel components.

The bridge rehabilitation was substantially completed in 2005; the entire bridge site was completed in 2007 (see Figure 13.4).

THE ARTHUR A. SMITH COVERED BRIDGE

DESCRIPTIONS OF STRUCTURE

The Arthur A. Smith Covered Bridge was built in 1870. The bridge structure is approximately 99 feet long by 17 feet wide with one 10-foot 11-inch wide lane. It is supported on a pair of stone masonry abutments. (See Figure 13.5.)

The structural system is comprised of a pair of Burr trusses, 9 floor beams, 7 longitudinal stringers, and transverse deck planks. Each Burr truss is formed with an eight panel timber truss, similar in configuration and function as a Howe truss, sandwiched between a pair of timber Burr arches.

The Burr truss has pairs of opposing timber diagonals arranged symmetrical about the centerline of the span. The interior diagonals are 6×8 timbers. The four end diagonals are 8×12's.

(b)

(c)

FIGURE 13.4 (continued)

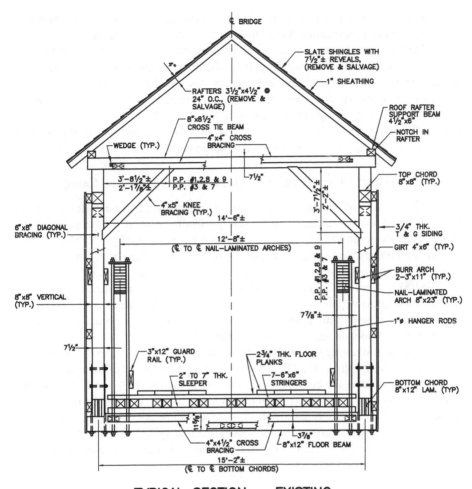

FIGURE 13.5   Cross section of existing bridge structure.

Vertical members are 8 × 8 timbers and are seated on top of the pair of 8" × 12" laminated timber bottom chords.

The floor beams are 8 × 12 timbers and are located at the bottom of the bottom chords supported by two steel hanger rods at each connection point. The pair of steel hanger rods at each connection point is attached with steel bolts to each face of the timber vertical members of the truss.

The stringers are 6 × 6 timbers that are supported on top of the floor beams. Transversely placed 3-inch thick timber planks provide the decking.

Each top chord of the truss is formed with 8 × 8's place between the 8 × 8 timber verticals. These top chord members are located approximately 18 inches below the top of the verticals.

On top of each verticals are 8 × 9 timber cross tie beams with 4 × 5 timber knee braces, forming the transverse bracing system to resist lateral wind load.

The lateral bracing system is provided by horizontal 4 × 4 cross braces between cross tie beams and horizontal 4 × 5 cross braces between floor beams.

Inside the pair of Burr trusses is a pair of nail-laminated timber compression arches, which provide additional support for the floor beams. Records are not available as to when the pair of compression arches was added to increase the structure's load carrying capacity. It is believed that the compression arches were added in the early 1900's. Given the age of this addition, the arches have become a part of the historic fabric of this bridge structure even though they weren't part of the original construction. The cross section of each nail-laminated arch is about 8" × 23".

DESIGN ISSUES AND MODIFICATIONS TO UPGRADE LOAD CARRYING CAPACITY

Although both timber structures suffered from neglect and severe deterioration from being exposed to the elements, the Arthur A. Smith Covered Bridge suffered the most. It had been removed from its abutments for a number of years. Many of the timber components have suffered severe dry rot or were broken from vandalism and carelessness during the structure's removal from its abutments. (See Figure 13.8.)

We had to carefully identify the physical conditions at each timber member to determine whether repair or replacement was in order. Dimensions of the overall framing have to be measured and extrapolated or interpolated in order to create the 3-D computer model as accurately as possible.

Unlike the Burkeville Covered Bridge, we determined that the existing structure of the Arthur A. Smith Covered Bridge could satisfy increased loading requirements due to the presence of the compression arches. (See Figures 13.6 and 13.7.) With the floor beams already tied to this pair of compression arches, it would be relative easy to modify each arch to carry the increased compression loads. This pair of compression arches provides in effect the secondary load carrying component, which the designer wanted in the rehabilitation design, with the necessary reserved capacity for potential overloads.

STRUCTURAL ANALYSIS

Similar to the Burkeville Covered Bridge, a GT STRUDL 3-D computer model was created to analyze and evaluate the structural framing system. The design loads and loading combinations were identical to the Burkeville bridge model. Location specific wind pressure and snow load according to the state building code were used.

Even though the arches attracted a significant portion of the design live load, many member components of the pair of Burr trusses were overstressed from the upgraded wind and live loads. Subsequently, many of the existing timber members had to be replaced with members that were slightly larger in cross section, using a compatible species of timber with higher design allowable stress values.

Since many of the existing components of the structure had to be replaced due to lack of capacity or had to be replaced because of defects due to deterioration, every attempt was made to reuse as much as possible the original timbers. We had specified shifting of low capacity diagonals, otherwise in good condition, to different locations of the truss where the available capacity satisfied the required load demand.

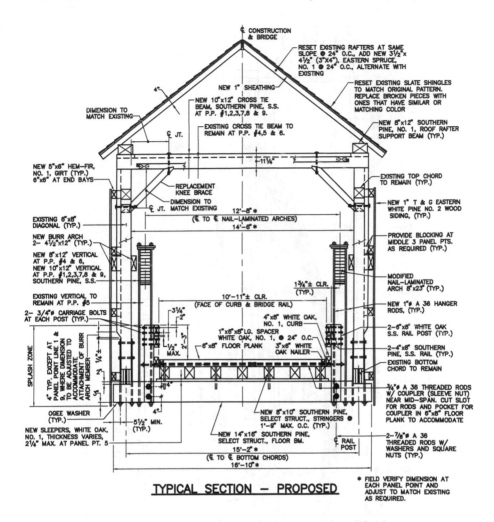

FIGURE 13.6 Cross section of bridge structure with proposed modifications.

In another case, we specified that any defective knee braces were to be replaced with timbers cut from salvaged original stringers. Due to the design load demand, all of the original stringers had to be replaced.

The bridge rehabilitation was substantially completed in 2006. The entire bridge site was completed in 2007. (See Figure 13.8.)

DISCUSSION AND CONCLUSIONS

The rehabilitation of historic timber covered bridges is a challenge for all project stakeholders. In the case of the Burkeville and Arthur A. Smith Bridges, stakeholders included the community, the local and state Historical Commissions, and the owner, MassHighway.

The covered bridges served as beloved infrastructure symbols for the communities, and town residents wished to have the bridges restored and back in service.

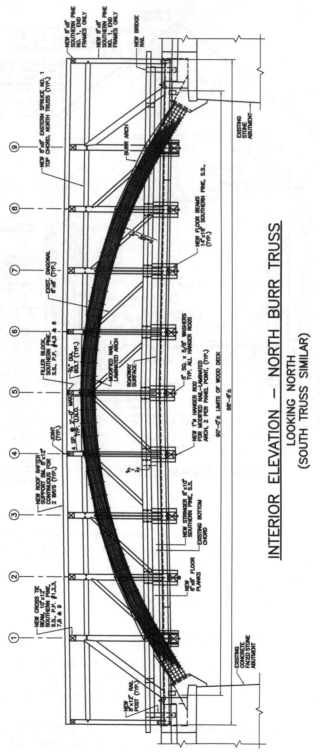

INTERIOR ELEVATION – NORTH BURR TRUSS

LOOKING NORTH
(SOUTH TRUSS SIMILAR)

FIGURE 13.7 Elevation of Burr truss with nail-laminated compression arch.

(a)

(b)

FIGURE 13.8 Arthur A. Smith covered bridge before rehabilitation (a) and after rehabilitation (b, c, d).

(c)

FIGURE 13.8 (continued)

MassHighway had to balance between the needs for public safety provided by the rehabilitated structures and associated costs. The Historical Commissions were concerned with preservation of the historic fabric of the structures and how much modification could be permitted before the bridges' historic value was lost. It was incumbent upon the designer to take into account all of these requirements, and develop rehabilitation plans that best satisfied all of the project goals.

The Burkeville Covered Bridge provides a good example of a situation where conflicting stakeholder requirements made the design difficult. Detailing for wind resistance where none had been provided before required structural details that initially were determined to be not in keeping with Historical Commission requirements. After discussions with MassHighway and time spent on various options to find a non-obtrusive detail, we finally settled on a supplemental sistering steel frame to provide the needed structural capacity for the upgrade.

Historic bridges provide benefits that are not so easily tabulated in a cost/benefit analysis. Covered bridges are structural icons representative of the New England countryside. In addition to their natural beauty, the bridges provide for a living museum of construction forms and the ways people lived in the past. When it is possible to rehabilitate the structures to meet current design criteria, then it is a case of "having our cake and eating it too": the bridges serve not only as rallying points for the local towns and as museums, but as actual functioning bridges. However,

(d)

FIGURE 13.8 (continued)

"eating the cake" can be expensive, more so than replacement with a modern but standard highway bridge. Also, at some point during the rehabilitation process, it can be argued that so much of the original function is lost that the bridge no longer qualifies as a historic structure.

A counterpoint to this argument is that the bridge, itself, is not static, but a structure that is amended and improved with time. For example, timber arches were added more than 50 years ago to increase the load carrying capacity of the Arthur A. Smith Covered Bridge. Even though the original construction didn't include these arches, over time they became accepted and appreciated as part of the overall historic fabric of the structure. Our efforts to rehabilitate both bridges will hopefully lead to their use and appreciation for many decades to come. Then, at some point in the future, improvements installed today may be looked upon in the same vein as the timber arches—as a part of the overall historic fabric of the bridges.

ACKNOWLEDGMENT

The authors express appreciation to the owner, MassHighway, and project participants for the opportunity to participate in this unique project.

14 The Historic Rehabilitation of the Market Street Bridge in Chattanooga, Tennessee

Ian C. Engstrom

The Market Street (Chief John Ross) Bridge in Chattanooga, Tennessee is a remarkable landmark situated as a major artery into the downtown area. Its status as the second longest double leaf bascule span in the United States is amplified by the graceful arches that lead up to the main span. The architectural relief present in the concrete is decorative and stately without being overwrought. The citizens of Chattanooga have embraced the bridge as an integral part of their history. The opportunity to participate in the historic renovation of such a structure presents itself only rarely and as such grants a unique occasion for today's engineers and craftsmen to test their talents in replicating the feats of earlier generations.

Prior to the Market Street Bridge there were two bridges constructed across the Tennessee River in Chattanooga; the military bridge, built during the Civil War, and the Walnut Street Bridge. The military bridge, a wooden arch structure with a drawbridge near the south bank, was the first permanent bridge in Chattanooga and was erected near the current Market Street Bridge location. Union soldiers constructed this bridge, which was promptly dubbed "Meigs' Folly" by the Army in dubious honor of the U.S. Quartermaster General due to its exorbitant cost of $750,000, in 1864, and left it to the City of Chattanooga when the army of occupation pulled out in 1866. By the following year the bridge was rendered useless when the greatest flood on local record reached its maximum stage of 57.9 feet, washing out portions of the already decaying structure. For the next 24 years the City of Chattanooga had to rely on a succession of ferry services to cross the Tennessee River.

In 1889 construction was begun on a river crossing approximately half a mile upstream at a high bluff on the south bank of the river. This location was chosen to eliminate the need for a movable span. The structure was completed in 1891 and became known as the Walnut Street Bridge. The bridge was designed to carry an electric streetcar line. In 1897 the bridge floor in one and a half spans was destroyed by fire. In 1911 an engineering report suggested that the timber deck needed replacing, and finally in 1914 the timber stringers were replaced with steel girders and a new timber floor was installed. These repairs were required in large part due to the increasing amount of automobile traffic in Chattanooga and the growth of the towns on the north bank of the Tennessee River.

While the Walnut Street Bridge had contributed greatly to the growth of Chattanooga and the surrounding towns, the increasing demands of the automobile traffic and the maintenance demands of the timber deck, not to mention the fire hazard, had resulted in the creation of a Tennessee River Bridge Commission by Hamilton County in January, 1914. The Commission was charged with letting a contract and supervising the construction of a new crossing at the foot of Market Street. The preferred material for the new bridge was to be concrete to reduce the maintenance concerns and fire hazard. A bond of $500,000 was provided for the bridge construction. In February, 1914, the design of B.H. Davis for a concrete bridge was provisionally selected by the commission. The site chosen by the commission called for a low level structure and the use of a movable span over the main channel. Mr. Davis specified a Scherzer type bascule for the main channel span and concrete arch span approaches. The conditional acceptance of this design was based upon some apprehension that the bridge as proposed could not be completed within the given bond appropriation. In order to get approval from the War Department, the main span of the bridge had to be revised to a length of 300 feet. The design was revised and approved in August, 1914.

Construction began in November, 1914. Exploratory borings taken in the river indicated bedrock of a limestone formation with numerous seams and fissures, which were filled with soft clay, at a depth of 22 feet below the proposed top of footings. The exception to this rule was on the north bank of the river where bedrock was located 50 feet below the proposed top of footings. The shore piers were constructed without any unusual difficulty. Pier 1 was able to be founded on bedrock within the limits suggested by the exploratory borings, and Pier 6 was founded on a spread footing due to the depth of bedrock on the north bank. Pier 2 was constructed with some difficulty due to problems with pumping out the cofferdam, but Piers 3 and 4 appear to have been constructed with no unusual problems. During the excavation for Pier 5 and the Pier 7 (the end pier for the north arch spans, referred to as the north abutment in the quote below), it was decided to carry the foundations to bedrock instead of using spread footings. At these two locations the foundations were constructed using the pneumatic caisson method as described below in the *Chattanooga News* from December 14, 1915:[*]

> The north abutment could not be treated in the same manner on account of the elevation of the springing line and the horizontal pressures due to the low rise in the crown of the arches. It was not found practical to trust this pier on a spread footer and it was decided to go to rock at about elevation minus thirty-five. Two holes were opened in this foundation 12×43 feet, timber sheeting being used to the elevation of the water.

---

[*] *Chattanooga News*, 14 December 1915.

FIGURE 14.1   Market Street Bridge, Span 2 under construction, 1915. (*Source:* Luken Holdings, Inc.)

An inside ring of steel sheeting was then driven to boulder formation and excavation was done by hand with the open caisson method to elevation minus fourteen. At this elevation the water was coming in at too great a volume to be handled in so small a cofferdam. It was then decided to use pneumatic construction. The water was pumped out until within about six feet of the bottom, a platform was inserted over the entire area of the hole, this platform was supported on 8×8 posts with 10×10 cross caps, using three-inch tongue and groove lumber for flooring. Two thirty-six-inch shafts were then placed vertically on the platform with the other piping, which conveyed the electric wiring and the air feed.

Sixteen feet of concrete was then placed in the hole, supported by the platform, after allowing a normal time for concrete to set, air was then turned on and work begun taking out platform and excavating. Rock was found at elevation minus thirty-six, and excavation was taken to a maximum of eighteen feet from the concrete block, which had been previously placed to serve as a bulkhead, the space then left between the rock and the concrete bulkhead was filled with cement under air pressure and when brought up to the bottom of the concrete bulkhead was packed firmly by hand and then grouted under 100 pounds of air pressure. This method of construction is unique and was very successful in every detail and found to be very much cheaper than the usual method of using caisson.

By September, 1915 there was optimism that the concrete elements of the bridge would be completed near the first part of 1916. The arch spans were complete up to Pier 2 from the south and the false work for Span 6 was being constructed. However, the north abutment and Piers 3 and 4 were still below the water line, and Pier 5 had just emerged from below the water line. On December 19, 1915, the arch span between

FIGURE 14.2   Market Street Bridge, under construction, 1916. (*Source:* Luken Holdings, Inc.)

Piers 2 and 3 was washed out by the combined effects of drift against the span false-work and high water. The damage set back construction about one month.

In July, 1916, the County Bridge Commission was called upon to defend itself against charges of malfeasance with regard to the rising cost of the bridge; it appears that, some time before this date, the commission had severed its relationship with Mr. Davis and hired Mr. J. E. Greiner to oversee the remaining design work. The commission laid the blame squarely on the shoulders of the design engineer, and the county unanimously gave a vote of confidence to the bridge commission. In spite of these various setbacks, the bridge opened to traffic in November, 1917. While the high quality of workmanship was uncontested, the final cost brought heavy criticism against the design engineer, Mr. Davis. The final cost of the bridge was $1.1 million, more than double the original bond appropriation of $500,000. There was much debate about where fault lay in the final cost of the bridge. Mr. Davis for his part filed suit against the county for breach of contract and eventually won a $15,000 judg-ment, in which he blamed the cost overrun on "unforeseen foundation conditions" and "unauthorized changes in construction ordered by the bridge commission and to delay due to foundation conditions, and to the action of the commission."*

FIGURE 14.3   Million Dollar Bridge over Tennessee River at Chattanooga, August, 1917. (Courtesy of the Library of Congress, LC-USZ62-123508.)

* *Chattanooga Times*, 7 May, 1920.

In its final configuration the Market Street Bridge is a sixteen span bridge with a total length of 2000'-0". The bridge consists of three concrete arch spans (165'-180'-180', clear span) on each side of the main 300'-0" (clear span) double-leaf, Scherzer type, steel truss, bascule span. The arch spans consist of a concrete deck on cast-in-place concrete frames transmitting loads to a single, solid rib arch member of varying depth. The superstructure is a closed spandrel wall system. There are nine, 40' (clear span) approach spans consisting of cast-in-place t-girders seated on column bents on the north end of the bridge. The parapet is an open railing with monuments and lighting pylons incorporated. There are numerous architectural corbels and relief details throughout the structure. At the time it was completed, the bascule span was the longest of its type in the world. The roadway was originally configured to carry two streetcar lines, as well as vehicular traffic.

With Chattanooga growing, and the Walnut Street Bridge in frequent need of repair, the Market Street Bridge bore much of the burden as the City's river crossing. The streetcar lines were removed sometime in the late 1930's. Increased demands were placed on the Market Street Bridge in 1948, when an engineering firm reported that the Walnut Street Bridge was "dangerous for heavy vehicles."* The bascule operating machinery, with its quirks, concerned locals over the years. Articles suggest that it was frequently out of service or in repair, and there was a three-year period noted in which the bascule was not opened for testing as required by Federal law. An article in the *Chattanooga Times* from October 24, 1948, states "The Market Street Bridge mechanism has been out of repair for some time, and some annual inspections by the government were discontinued because it was feared the weight of the massive center steel beams might cause them to give way if lifted for a test."† Plans for rehabilitation began that same year, and were completed in 1949. The rehabilitation included mechanical and electrical restorations, replacing portions of the wooden sidewalk, cleaning and painting the bascule span, and replacing the original timber roadway on the bascule span with an open steel grid floor.

An article in the *Chattanooga Times* from January 4, 1950, stated that after urging from the Volunteer Chapter of the Daughters of 1812, the City Commission approved changing the name of the Market Street Bridge to the Chief John Ross Bridge to commemorate the origins of Chattanooga.‡ In 1959, the US Highway 27 bridge was constructed about half a mile downstream and the strain on the Market Street Bridge was relieved. Discussions about replacing the bridge started as early as 1973 when an article in the *Chattanooga Times* doomed the bridge to demolition stating, "The Market Street Bridge lift mechanism, a plague upon the City's house for decades is inoperable because of deterioration and specifications are being drawn in preparation for taking bids to replace it."§ The bridge was unable to be raised upon request by a passing vessel during high water in March, 1973. Movements to rehabilitate the bridge began in 1978 when local agencies took up the cause to preserve the bridge's structural integrity and restore its original visual character. The City supported these plans, but they never came to fruition.

---

* *Chattanooga Times*, 14 May, 1948.
† *Chattanooga Times*, 24 October, 1948.
‡ *Chattanooga Times*, 4 January, 1950.
§ *Chattanooga Times*, 12 April, 1973.

FIGURE 14.4   Market Street Bridge, May, 2000. (*Source:* Ian Engstrom, Parsons Transportation Group Inc.)

At that time, the counterweights were painted with advertisements and freeway style lights had long since replaced the historic light fixtures once adorning the pylons and monuments. While some thought that the Market Street Bridge was doomed for replacement, others proclaimed that the bridge was structurally sound. In 1985, a *Chattanooga Times* article quotes Public Works Commissioner Paul Clark stating "I've been down to every inspection on that bridge and I don't know where the idea came that the bridge is unsafe."[*]As ideas for riverfront redevelopment surfaced, in 1986 the *Chattanooga Times* predicted that the bridge would be spared from being demolished and stated that it "will serve this city for many years to come as the major traffic entrance into a redeveloped urban waterfront and the heart of downtown."[†]The State of Tennessee decided to give the bridge a major facelift by repairing concrete, removing the advertising on the counterweights, replacing some deteriorated steel, repainting the bascule span, and repaving the roadway. Although some of the structural issues were being addressed, the repairs were an unsightly mess with little thought given to the aesthetics of the bridge. Mayor Gene Roberts told members of the Chattanooga Engineers' Club that the bridge "looks like just some old pair of patched pants."[‡]

The Tennessee Department of Transportation was concerned with upgrading the traffic safety features of the bridge while maintaining its unique architectural and historic features. To that end, Parsons Transportation Group was retained by the Tennessee Department of Transportation in May, 2000 to do a full inspection and prepare recommendations as the first phase of the rehabilitation and historical renovation of the Market Street Bridge. An extensive inspection process documented the condition of the bridge. This inspection included the installation of access hatches in the sides of the spandrel walls to inspect the hidden superstructure, which had not been inspected in many years. Destructive testing, including concrete cores and steel coupon samples, was utilized to determine the properties of the concrete for rating the existing structure and whether or not welding could be utilized in the repair methods for the steel structure. Non-destructive testing was utilized to determine the reinforcing of existing members, which would also be useful in rating the existing structure, where the existing plans were unclear or just not legible.

There were numerous instances of cracking, spalling, and exposed reinforcing throughout the concrete superstructure. The freeze/thaw cycle appeared to be the primary mechanism of concrete deterioration. The architectural elements of the concrete structure suffered the worst and showed heavy spalling, cracking, and efflorescence. There was little or no reinforcing to control cracking in these architectural

---

* *Chattanooga News-Free Press*, 25 June, 1985.

† *Chattanooga Times*, 16 April, 1986.

‡ *Chattanooga Times*, 25 November, 1986.

FIGURE 14.5   Deterioration at Abutment 1, 2000. (*Source:* Ian Engstrom, Parsons Transportation Group Inc.)

elements. Several instances of improperly supported and tied reinforcing steel were found. This condition manifested itself as exposed steel at the face of internal arch members and as spalls where inadequate cover fell away. Thermal movement of the structure had caused internal members of the superstructure frames in the arch spans to crack as well as the north abutment of the approach spans. The spandrel walls were cracked along the top of arch member, suggesting stress from thermal movement. Expansion joints were clogged with debris and were not functioning, which induced further stresses in the north approach spans. The piers exhibited cracks above and below the water line. There was little or no inspection history available for these cracks. From the weathering and the lack of apparent tributary effects in the rest of the structure, these cracks were estimated to be very old and not structural in nature. The underwater inspection of the pier foundations revealed

FIGURE 14.6 Deterioration of Arch Superstructure Frame, 2000. (*Source:* Ian Engstrom, Parsons Transportation Group Inc.)

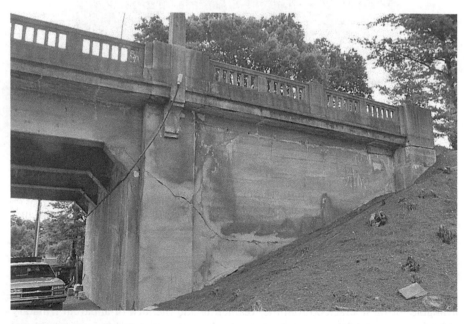

FIGURE 14.7 Deterioration at Abutment 2, 2000. (*Source:* Ian Engstrom, Parsons Transportation Group Inc.)

FIGURE 14.8   Pier 4, 2000. (*Source:* Ian Engstrom, Parsons Transportation Group Inc.)

some unusual configurations, which prompted a visit to the local library. It was this visit that yielded the newspaper clippings, which described the construction of the bridge in some detail, and helped to understand the difficulties encountered during construction.

The bascule span had several deteriorated members in areas where it is typical to find such deterioration in a steel truss span from accumulated debris and bird droppings, and the track girder, which supports part of the bascule mechanism, on one of the fixed rack frames had buckled. This failure appears to have occurred some time ago and has had little impact on the operation of the span other than some uneven wear in the bearings as was discovered later. The timber decking of the sidewalk on the bascule span was also due for replacement. The decking exhibited checking and splitting, and there was deterioration of the steel stringers under the timber nailers.

The original recommendations detailed below for the rehabilitation and renovation of the bridge are intended to be representative of the kinds of repairs that were to be implemented and is not a complete list of the repairs recommended for the

FIGURE 14.9 Fixed Rack, Buckled Track Girder. (*Source:* Ian Engstrom, Parsons Transportation Group Inc.)

structure. The bridge parapet and overhang were to be replaced for the length of the concrete structure. A prototype of the bridge parapet was to be crash-tested to the Level 1 criteria in accordance with NCHRP Report 350. Deteriorated members of the bascule span were to be replaced as well as the timber sidewalk. The architectural fascia on the arch members was to be removed to sound concrete and a new fascia, which was reinforced and anchored into the arch member, constructed. Surface repairs were to be made to the underside of the arch members. Surface repairs, which included concrete repair and epoxy injection of cracks, were to be made to the faces of Piers 1 through 6 up from the waterline. The approach spans were to have extensive repairs including parapet replacement, beam replacement or repair, and surface repairs to the substructure units. The entire concrete surface of the bridge above the waterline was to be cleaned and receive a painted finish to match new and existing concrete. Underwater repairs included the pouring of a concrete seal over the existing exposed footings and epoxy injection of cracks. The mechanical works were to be updated and overhauled by replacing the existing motors and restoring manual span operation in the event of motor or electrical failure.

Every attempt was made in the design phase to conform to standard DOT specifications and standardize repair details. Standard repair procedures and materials are important in a project of this type in order to keep costs down. The standard DOT specifications were deemed adequate for this application, and there was little or no need to add new repair procedures and complexity to an already complex project. Architectural details were categorized into as few types as possible in order to facilitate the reuse of concrete forms. Standard expansion joint details were utilized in the north approach spans to improve the function of the bridge and make future maintenance easier. Precast elements were utilized wherever possible, from the prestressed girders and deck panels of the north approach spans to the lighting pylons that were incorporated into the parapet. The replication of the bridge parapet and arch fascia were the most difficult aspects of the concrete repairs, requiring the most effort to measure the existing details, design the replacement, and in the end construct the new elements.

One of the specifications that did require some adjustment was the concrete finishing. The completed bridge was going to be a patchwork of old and new concrete, which was the same problem that had plagued earlier repair projects for this structure. The standard concrete finishing specification included a coarse texture finish coating, which is appropriate to normal highway structures, but would obliterate the architectural details under renovation. A finish coating was chosen by the architect with enough texture to blend the old and new sections but not to cover up details. A test section was also required in the special provision to verify the assumptions. During construction, the painting subcontractor recommended a stain product for application to the bridge parapet. This was found to be an acceptable and in some ways a better product, since the entire bridge parapet was new construction and there were minimal issues of blending at the four original monuments, which stood at the beginning of the arch spans on the north and south sides.

This original list of repairs from June, 2001, while consistent with the Department's original intent for the structure, was modified over the course of the project based upon input from local agencies. The City requested that the sidewalks be widened, on the concrete portion of the bridge, by 3 feet on each side to improve pedestrian access. The effect of this widening structurally was that the existing exterior superstructure members (spandrel walls or girders, depending on the location) needed to be replaced to handle the increased torsional loading. The City also had a great deal of input on the selection of a light fixture to be used on the bridge. It was also requested that the project be delayed for two years while the City completed a riverfront development project in the vicinity of the bridge. Upon determining that there were no critically deficient items that required immediate attention, this delay was granted and construction did not begin until September, 2005.

For the duration of construction the bridge was closed to traffic for two reasons. The width of the roadway (36 feet) was not sufficient to allow for more than one lane of traffic on the bridge during construction, and it was determined that the time required and the construction costs would increase substantially if the construction work was required to be phased. The project was set up to be completed

FIGURE 14.10 North Approach Spans, Demolition, 2005. (*Source:* Ian Engstrom, Parsons Transportation Group Inc.)

in two years with incentives for early completion and disincentives otherwise. The winning bid for construction was awarded to Mountain States Contractors, LLC for approximately $13 million.

During construction there were a few issues that changed the scope of the work to be performed. While in the process of removing the approach spans superstructure, two columns were damaged by the contractor and were visibly out of plumb. In excavating around the damaged columns it was discovered that the connection between the column and the footing was seriously deteriorated. Further investigation of unaffected columns revealed similar deterioration at the column-footing connection. At this point it was decided to replace all of the existing columns in the approach spans. Additional investigation of the north abutment revealed a hidden span behind the abutment wing walls, where fill had been assumed as the existing plans were unclear. Here the condition of the columns and beams also called for the removal and replacement.

The overhaul of the mechanical works in the bascule span revealed wear in the large hollow shafts that ran from the cog on the fixed rack frame to the first flywheel in the gear system. This wear was not visible until the system had been taken apart to replace the bearings. One shaft had to be replaced entirely, and the others had to be built up and milled to the correct diameter.

The repair work to seal the footings at the four river piers uncovered issues with foundation stratum that made the cofferdam work more difficult than originally

FIGURE 14.11   North Approach Spans, Reconstruction, 2005. (*Source:* Ian Engstrom, Parsons Transportation Group Inc.)

thought. At Pier 3, the original scope of work involved excavating down to bedrock, setting a cofferdam around the pier and pouring a new tremie footing that sealed and covered the original footing. During excavation and setting of the cofferdam ring, rock outcrops were discovered that made required resizing the cofferdam to ensure a good seal against the bedrock. The larger cofferdam would have encroached into the navigation channel by an unacceptable amount. The repairs to Piers 3 and 4 below the waterline were then reclassified to be done as underwater repairs by a construction diving service. In this way voids uncovered by the excavation process and cracks in the pier were sealed underwater by divers. The cracks below the water line were sealed with a chemical foam grout instead of a structural epoxy grout. As was mentioned before, these cracks were determined not to be a structural issue, and the injection of these cracks was performed to seal the pier against water intrusion and future deterioration. The correction of these minor issues is of long term importance to prevent future deterioration and to remove these items from future inspection reports, where they might cause undue concern to an inspection team that is not familiar with the structure and its history.

The construction was substantially complete on July 28, 2007 and the bridge was reopened to traffic on August 5, 2007, which was 45 days ahead of schedule. The City celebrated the reopening with a weekend-long celebration that gave the public a chance to give the bridge a close-up inspection before it was opened to traffic.

FIGURE 14.12 North Approach Spans, Reconstruction, 2005. (*Source:* Ian Engstrom, Parsons Transportation Group Inc.)

FIGURE 14.13 Mechanical System, 2007. (*Source:* Ian Engstrom, Parsons Transportation Group Inc.)

FIGURE 14.14   Mechanical System, 2007. (*Source:* Ian Engstrom, Parsons Transportation Group Inc.)

FIGURE 14.15   Pier 2, Cofferdam, 2006. (*Source:* Ian Engstrom, Parsons Transportation Group Inc.)

FIGURE 14.16 Market Street Bridge, Bascule Span, August, 2007. (*Source:* Ian Engstrom, Parsons Transportation Group Inc.)

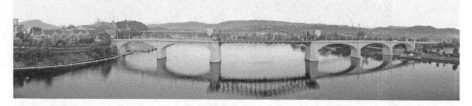

FIGURE 14.17 Market Street Bridge, September, 2007. (*Source:* Ian Engstrom, Parsons Transportation Group Inc.)

Throughout the project the Parsons team variously consisted of design engineers in the Memphis [structural design, construction engineering and inspection] and Chicago [structural, mechanical and electrical design] offices, and sub-consultants including Volkert & Associates [roadway design, construction inspection], Franklin Associates Architects, Electrical and Electronic Controls [bascule span operation], MACTEC Engineering & Consulting, Inc. [destructive and non-destructive testing and sampling of original bridge members], K.S. Ware Associates [materials testing and construction inspection], and Mainstream Commercial Divers [underwater inspection services].

REFERENCES

Gaston, Kay. "Walnut Street Bridge." Historic American Engineering Record 1979. Ed. Donald C. Jackson and Jean P. Yearby. HAER, 1985. HAER No. TN-11.
"Work Starts Tomorrow On New County Bridge." *The Chattanooga Times*, November 15, 1914.
"New Record In Concrete." *Times*, March 2, 1915.
"New Bridge In 90 Days." *Times*, September 18, 1915.
"...Being Constructed By Vang Co." *The Chattanooga News*, December 14, 1915.
"Bridge Span Disappears." *Times*, December 20, 1915.
"Four Weeks For Repairs." *Times*, December 21, 1915.

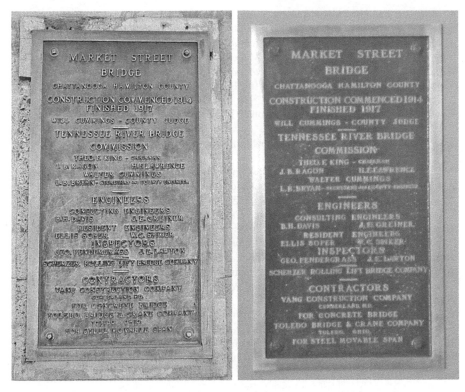

FIGURE 14.18   Original and Restored Dedication Plaques. (*Source:* Ian Engstrom, Parsons Transportation Group Inc.)

"Hot Reply To Investigators." *Times*, July 28, 1916.
"Down Comes Big Cable At Market Street Bridge." *Times*, March 1, 1917.
"Paving Of New Bridge Next Task of Commission." *Times*, July 11, 1917.
"Lights For New Bridge." *Times*, July 24, 1917.
"Final Test Of Bascule." *Times*, August 3, 1917.
"Big Bascule Stands Test." *Times*, August 4, 1917.
"Bridge Cost In Figures." *Times*, September 16, 1917.
"Bridge Done In Two Weeks." *Times*, October 7, 1917.
"Bridge Cost Reasonable." *Times*, October 24, 1917.
"New Bridge Is Dedicated." *Times*, November 18, 1917.
"Davis Tells Bridge Story." *Times*, April 28, 1920.
"Bridge Man Backs Davis." *Times*, April 30, 1920.
"Heavy Verdict For Engineer." *Times*, May 7, 1920.
"State Leaders Act To Expand Bridge Access." *Times*, May 14, 1948.
"Guide Lights To Be Put On Market Bridge." *The Chattanooga Free Press*, June 3, 1948.
"State Survey Set On Market Bridge." *Times*, October 24, 1948.
"Market Bridge Repairs Voted; $133,200 Cost." *Free Press*, January, 31 1949.
"Market Bridge Ready To Open." *Times*, December 15, 1949.
"Span Is Renamed 'Chief John Ross'." *Times*, January 4, 1950.
"Market Street Bridge Elevated By Accident, Closed For 6 Hours; No One Injured, 2 Cars Damaged, Blame Not Fixed." *Times*, April 8, 1950.

"Lifts Open Suddenly When Switch Closed During Greasing Job." *Free Press*, April 7, 1950.

"Bridge Lift Apparatus To Be Replaced." *Times*, April 12, 1973.

"Bridge Closed Part Of Day While State Makes Repairs." *Times*, April 12, 1977.

"Sabotage Shuts Market Span." *The Chattanooga News-Free Press*, July 5, 1977.

"Ross Bridge Controls 'Hot-Wired'; Traffic Rerouted For Several Hours." *Times*, July 6, 1977.

"Footprints Indicate 2 Persons Were Involved In Tampering With John Ross Bridge Mechanism." *Times*, September 7, 1977.

"Chattanooga's 'Tolerance Tester'—The Market Street Bridge Traffic." *News-Free Press*, May 20, 1978.

"Market Span Crumbly But Safe." *Times*, August 27, 1978.

"Market Street Bridge Was Largest Of Its Kind." *Times*, August 27, 1978.

"City Backs Restoration Of Bridge." *News-Free Press*, November 7, 1978.

"Commission Plans Bridge Rehabilitation." *Times*, November 8, 1978.

"Barge Crash May Spur Action For New Bridge." *News-Free Press*, November 24, 1978.

"Asphalt At Bridge Stripped To Provide Grooves For Safety." *Times*, April 4, 1980.

"Restore The Chief John Ross Bridge." *Times*, April 3, 1984.

"Improvement Is Not Impossible." *Times*, September 7, 1984.

"Market Street Bridge Back To 2-Lane Use." *News-Free Press*, March 26, 1985.

"Attention Needed." *Times*, May 29, 1985.

"Clark Criticizes Plan To Replace Bridge." *News-Free Press*, June 25, 1985.

"Sprucing Up The Market Street Bridge." *Times*, April 16, 1986.

"John Ross Bridge Gets Festive Lighting Soon." *News-Free Press*, January 28, 1993.

"Market Street Bridge Renovations Delayed Until 2005." *The Chattanooga Times-Free Press*, March 4, 2005.

15 Reinventing Squire Whipple's Bridge

Joseph J. Fonzi and Preston Vineyard

CONTENTS

ABSTRACT

Buffalo's new Commercial Slip Bridge is a modern interpretation of the original Whipple Arch Truss Bridge that once stood at the same location. Patented in 1841 by Squire Whipple, the Whipple Arch Truss was used extensively along the Erie Canal. The new, 100-foot span pedestrian bridge has become a signature piece for Buffalo's new inner harbor redevelopment. This paper discusses the design and construction of the modern interpretation of the Whipple Arch Truss Bridge and compares the modern design with the original patented design.

HISTORY

The construction of the Erie Canal in 1825 established an all-water passage from the Great Lakes through Buffalo to the port of New York City and the world, transforming Buffalo from a frontier village into a thriving commercial and industrial metropolis. For most of the 1800s, Buffalo's canal district stood at the center of the city's growth and development, the fulcrum between the Erie Canal and Lake Erie. The Commercial Slip and Central Wharf were key locations. Influenced by the flow of goods, people and ideas, a port culture emerged among the slips, wharves, grain elevators, warehouses, businesses, saloons, shops, residences and hotels of the canal district that shaped the character of the city.

FIGURE 15.1   Circa 1870 Commercial Slip Bridge photo.

The Whipple Arch Truss, which was built from a design patented in 1841 by Squire Whipple, spanned Buffalo's Commercial Slip. One of the most famous engineers of his time, Whipple was the first person to understand the stresses in truss members and he developed the theoretical formulae to calculate stresses in an articulated truss. His truss was the first to use cast iron for compression members and wrought iron for tension members. The design was so well regarded that the State of New York later accepted the bridge as their official standard design. It was used extensively on crossings along the Erie Canal.

The Commercial Slip Whipple Truss boasted three arches, each with nine panels. The arches separated the two traffic lanes and the two outboard pedestrian walkways. Based on standardized drawings for the Whipple Arch-Truss Bridge, all indications suggest that it spanned 100 feet.

This Whipple Truss Commercial Slip crossing was later replaced with a railroad bridge and the slip, as with most of the canal, became a storm sewer. The railroads ended the canal era and the use of the unique, mass produced Whipple arch truss.

DESIGN

Empire State Development Corporation (ESDC) commissioned Parsons Brincker-hoff, Inc (PB) to recreate the commercial slip and design the abutments for a modern prefabricated bowstring pedestrian bridge. During the schematic design phase, the project's architectural team requested that an actual Whipple truss recreation or interpretation be designed and locally fabricated. However, through our work on the project's master plan and design report, extensive research had been conducted of the commercial slip including the bridge; knowing that an exact replica using cast iron for the arches and timber beams would not be cost effective for the project's six year old construction budget, interpretive re-creation using modern materials was decided upon. PB held value engineering sessions with the construction manager (LP Ciminelli) and ESDC at 30% and 60% design to help control costs and

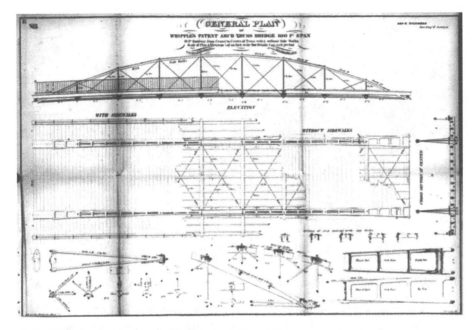

FIGURE 15.2   100-foot span Whipple Arch Truss Bridge patent drawing.

assess constructability. In the end, the bridge construction contract was awarded at a price below the programmed budget and the bridge was built with no significant changes to the construction documents, proving that the value engineering sessions were successful.

Using photographs of the original bridge and the patent drawings, accurate geometry of the structure was re-created. A 3-D CAD solid model was then developed to facilitate detailing and investigate constructability throughout the design. The following sections describe the various parts of the bridge, discussing Whipple's original patent design and the modern interpretation.

THE ARCHES

The original patent design's arches consisted of cast iron segments that overlapped at the hanger rod opening and butted together at the outer portions. The arch segments relied on compression to force the segments to bear on one another similar to a stone arch. The cross-section of the ribs was essentially angles with a seven inch vertical leg and a four inch horizontal leg. These could be easily replicated in the modern design by using standard structural steel angles of similar size. In the original design, the transverse bracing connecting the arch ribs together was a cast tee shape. The modern interpretation utilized a standard structural steel angle welded to the arch rib.

To improve lateral stability it was decided to connect the arch segments rather than letting compression alone hold the arches together. The detailing of this connection would be critical to the aesthetic of the arch as well as its constructability. The first concept was a bolted connection, possibly using a lap splice or fish plate

FIGURE 15.3   Proposed Whipple Arch Truss design.

type connection. Further study of the bolted connection revealed that it would not be feasible to assemble due to interferences with the hanger rods and cross bracing connections. A groove weld at the intersection of the arch segments was selected during final design. In addition, the segments were linked together using the hanger rod bearing plates and cross bracing connection plates welded to the ribs.

THE DECK

The preliminary design concept utilized rectangular hollow structural shapes to resemble the original bridge's timber beams. This concept was changed to wide flange shapes during the value engineering process because it was determined that the wide flange section would be more cost effective and easier to connect when compared to the hollow rectangular section. Changing to wide flange shapes required changing the hanger and tension rods connection details, which are described in a later section.

The decking is wood similar in design to the original structure; pressure treated yellow pine was selected for durability and longevity. Rough sawn timbers of similar size to the originals were hosen. These three inch by eight inch timbers are attached to the wide flange beams with galvanized carriage bolts. The top of the bolts form an aesthetic pattern on the deck. Carriage bolts were chosen for their ease of installation and future replacement.

In Whipple's original design, the openings in the deck under the arches were closed with two timber planks forming a gable roof-like cap, with the ridge on the arch's centerline. The boards were notched at the ridge to fit around the hanger and diagonal rods. The modern interpretation design uses bar grating. The grating was placed flush with the wood decking so pedestrians could walk freely under the arches. The open bar grating provided the lighting designer with a space for up-lighting to illuminate the arch at night and gives pedestrians a view down to the water and a portion of the bridge structure.

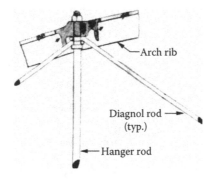

Arch rib

Diagnol rod → (typ.)

← Hanger rod

FIGURE 15.4   Upper connection original patent design.

TENSION MEMBERS

The hanger rods in Whipple's original design were wrought iron, threaded on both ends with an upset collar below the threading on the top just below the bottom of the arch. This was a stop or bearing for the diagonal rods. The center arch segment used an upside down "Y" hanger rod, the single end passed through the arch and the split end passed through the beam hanger castings called connecting blocks by Whipple. This split would give the center of the arch added transverse stability. One side of the "Y" would resist in tension as the top of the arch deflects or buckles transversely. The hanger rods used in the interpretation do not have a forged upset for the diagonal rods and are made of galvanized rods threaded at both ends. The "Y" hanger was split into two discrete rods that are angled out similar to the original design. Spherical washers were used to keep the nuts aligned with the rods. When the deck floor beams were changed from hollow rectangular to wide flange shapes, the hanger rod detail changed to four rods per hanger, or panel point, with two rods on each intersecting truss segment straddling the deck floor beam's web below.

The thrust tie or bottom chord of Whipple's original truss used closed iron links looped over the connecting blocks fixing the distance between the lower panel points. The end links were open ended and passed though the end arch segment's "foot", with a nut to secure each end of the "U" link to the arch foot. Tightening these nuts brought the arch ends together, compressing the arch and bringing it into proper geometry. The thrust tie on the interpreted design consists of four continuous galvanized threaded rebar tie rods per arch. A plate with a short segment of pipe is welded to the bottom of the deck floor beams to act as a guide and support for the tie rods in addition to hiding the splice coupling. The tie rods pass through the arch feet and are anchored using spherical nuts on a countersunk plate.

The diagonal rods are attached to the deck floor beams with a clevis attached to a plate in the center of the four hanger rods at each panel point. The top connection was similar to the bottom using a plate welded to the arch sections and a clevis on the rod. The rods and clevises are right hand and left hand threaded so the rods can be tightened without using a turnbuckle. The diagonals in Whipple's original design are somewhat different. In the original design, the diagonal rods passed through the hanger casting and tightened with a nut at the bottom. At the arch connection, the diagonal rod had a forged eye through which the main hanger rod passed. The eyes

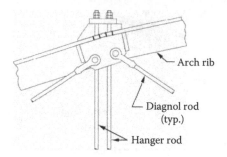

FIGURE 15.5   Upper connection detail modern interpretation design.

rest on an upset portion of the hanger rod and are tightened against the arch with the top nut of the hanger rod.

HANDRAIL

In conjunction with the landscape architect, the handrail was the last element to be designed. The architectural team requested that the bridge's handrail and site's handrail be similar in a way that would tie the site together. Photographs indicated that the original handrail was likely wrought iron, picket type with a spring scroll at the posts. The curve of the spring scroll was found to be a feature that would add aesthetic interest, and as in the original use of the scroll, it adds strength to the post against horizontal load. To give the bridge a finished edge, the top rail was designed using a wood dowel to match the rest of the site. The handrail is sectional consisting of two sections per arch segment.

ANALYSIS

TOP CHORD BUCKLING

The traditional design approach to address the buckling problem of the compressive top chord of a pony truss, or half through truss, is to consider the compressive chord a column that is braced by elastic lateral restraints at the truss panel points. The stiffness of the elastic lateral restraints coincides with the stiffness of the truss transverse frame. The critical buckling load for the member between elastic supports is then calculated taking into account the equivalent column slenderness ratio.

   The transverse behavior of the Whipple Arch Truss is actually quite different than a typical through truss structure. The compression chord of the Whipple Arch Truss is not laterally braced by elastic restraints since the vertical members of the truss are comprised of hanger rods that have negligible lateral stiffness. The Whipple Arch Truss obtains its lateral stability by the frame action that occurs as a result of bifurcating the compression chord so that the chord cross-section is wider at the abutments than at the mid span (Figure 15.7). Whipple's description of the lateral stiffness in the original patent reads, "It will readily be seen from the manner of forming and connecting the segments, that the top of the arch will present the appearance of two horizontal arches touching one another in the middle and

FIGURE 15.6 Photo of bottom chord connection.

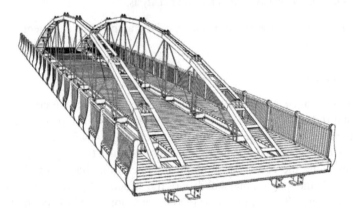

FIGURE 15.7 Isometric view showing bifurcated arch rib.

curving equally therefrom in opposite direction to the ends. This form being given it to produce stiffness and sustain it against lateral flexure." Whipple's arch rib design actually resembles and behaves similar to a Vierendeel truss, which had not been invented at the time the Whipple Arch Truss was patented. In addition to the frame action, the hanger rods and diagonals engage as the arch ribs displace laterally and provide additional resistance to further defection.

According to the original patent for the Whipple Arch Truss, the width of the chord cross-section, or rib, at the end of the arch should be approximately three times the width of the rib at the midspan.

Similar to the original patent design, the interpreted design's arch ribs are bifurcated and have a width that is three times greater at the end of the arch than at the midspan. To verify the stability of the arch rib, a 3-D finite element computer program was utilized to conduct a nonlinear buckling analysis (Figure 15.8). The nonlinear buckling analysis was performed using the iterative full Newton-Raphson method, utilizing incremental loads to determine the buckling behavior of the structure. The

FIGURE 15.8   Deflected shape from buckling analysis.

analysis concluded that the lateral capacity of the arch ribs is sufficient to withstand design loads.

Diagonal Members

Similar in appearance to Whipple's patented design, the modern interpretation design has diagonal rods that connect each panel point in the vertical plane of the arch rib. The diagonal members of the modern design behave somewhat differently and play a less important role than in the original design. Because the original design had pin connections along the compression chord members, the diagonal rods were relied upon to provide the longitudinal stability of the structure by preventing shear deformation of the truss. The segments of the compression chord of the modern design are welded together, forming a single member that has the capacity to resist shear deformation and bending. Although the diagonals do carry some tension and help reduce the bending moment in the arch rib, the modern design does not rely solely on the rods for the structures longitudinal stability. The diagonal members of the modern interpretation design add significant value to the aesthetics of the structure and results in a bridge that more closely resembles Whipple's original design.

CONSTRUCTION

The contractor elected to fully assemble the bridge, except the wood decking, on a barge upriver from the construction site. The arch ribs were shop welded into halves to facilitate shipping. The arch halves and beams were delivered to the erector's site painted and ready for assembly. The steel deck grillage was assembled on blocking to allow for the hanger rods to be installed without load and to easily adjust the arch geometry. The camber of the bridge was controlled by adjusting the length of the thrust tie. The bridge was then tugged into the slip with the City's fire boat, the *Edward M. Cotter*, leading the way and breaking the ice. A 500 ton hydraulic crane

FIGURE 15.9   Lifting the bridge into its final position.

FIGURE 15.10   Completed structure.

lifted the bridge into position onto the abutments. The anchor bolts and the bridge foot holes lined up well within tolerance. Throughout the winter the wood decking and bar grating were installed. The contractor used a template to drill the bolt holes through the wood and the steel flanges to install the carriage bolts. The ice was thick enough that an iron worker was able to work on the underside of the bridge from an A frame ladder sitting on the piece of road plate on the ice (with proper tie offs).

CONCLUSION

Today, the bridge is almost complete with a final walk through and punch list to take place shortly. ESDC was very pleased that a custom historic replica was built within budget, and the bridge has become a signature piece for Buffalo's new inner harbor. More than simply connecting two pieces of land bound by water, it's a bridge between the City's proud, storied past and its present efforts to rebuild the harbor in support of commercial and residential development.